C++ and C# programming for beginners

Crash Course fprogram to learn from scratch C++ and C# languages. Develop new coding skills with hands on projects.

Michail Kölling

© **Copyright 2020 - Michail Kölling - All rights reserved.**

The content contained within this book may not be reproduced, duplicated or transmitted without direct written permission from the author or the publisher.

Under no circumstances will any blame or legal responsibility be held against the publisher, or author, for any damages, reparation, or monetary loss due to the information contained within this book. Either directly or indirectly.

Legal Notice:
This book is copyright protected. This book is only for personal use. You cannot amend, distribute, sell, use, quote or paraphrase any part, or the content within this book, without the consent of the author or publisher.

Disclaimer Notice:
Please note the information contained within this document is for educational and entertainment purposes only. All effort has been executed to present accurate, up to date, and reliable, complete information. No warranties of any kind are declared or implied. Readers acknowledge that the author is not engaging in the rendering of legal, financial, medical or professional advice. The content within this book has been derived from various sources. Please consult a licensed professional before attempting any techniques outlined in this book.

By reading this document, the reader agrees that under no circumstances is the author responsible for any losses, direct or indirect, which are incurred as a result of the use of the information contained within this document, including, but not limited to, — errors, omissions, or inaccuracies.

Michail Kölling

Table of Contents

C++ Programming

Introduction 7

Anatomy of C++ 9

Declaring constants 21

Functions 31

Polymorphism 35

Operator type and overloading 43

Macros and template 51

Classes 65

Library 81

STL 91

Conclusion 95

C# Programming

Introduction 99

Anatomy of C# 101

Data Type 119

Operators Variable 123

String and list 129

Syntax 131

Classes 135

LINQ, queries, operators — XAML 143

Program to make decision 151

Net 155

ENUM and struct 159

Common mistake and how to avoid them 163

Conclusion 169

C++

IV

C++ PROGRAMMING

After work guide to master C++ on your own. Build your coding skills and learn how to solve common problems. Transform your passion in a possible job career as a computer programmer.

Michail Kölling

C++

Introduction

Welcome to the wonderful world of programming, the chapters contained in this book will give you a basic understanding of programming in C++. By its final chapter, you will be able to create a complete program on your own, using C++.

This guide is aimed at newcomers to C++. If, however, you are entirely new to programming, I recommend first reading our primer to programming. It covers all the concepts, terms, programming paradigms, and coding techniques that a complete novice needs to know.

This guidebook is going to take some time to look over all of the things that we can do with the C++ language, and how beneficial learning this language can be.

This object-oriented programming course in C++ language presents learners with the concepts and techniques necessary to design, develop and implement a robust program model effectively. As a learner, you will be able to grasp practical knowledge on how to apply the fundamental concepts of object-oriented analysis and design and solve various problems in your day to day activities.

C++ is a computer programming language widely used for general-purpose programming. It is an extension of C-language. The basic understanding of C++ can be acquired from C. That's why both computer languages are represented as C/C++. Bjarne Stroustrup developed this multi-paradigm language in 1979.

C++

In today's world, many operating systems use C++ as their basic language. Some system drivers, browsers, and games are based on C++ programs. It is a free form, compiled, and statically typed programming language. Many professionals believe that C++ is the most efficient language to achieve the desired results.

Learning code is ultimately the language of the future. We have all heard something close to that at some point in our lives.

By the time you reach the end, you should have no problem reading C++ code and writing programs that are both interesting and useful. So, let's dive in and learn C++.

Chapter 1

Anatomy of C++

Introduction to programming languages
As you already know, the program is nothing but a set of instructions. These instructions are executed by the hardware, which is the physical computer machinery. Though modern computers are fast, they have their limitations. Computers can only understand a set of instructions given in their native language. For this, you should understand the concepts of machine language, assembly language, and high-level languages.

Machine language
Though computers are very advanced machines, they cannot understand languages like C++ directly. A computer only understands 0s and 1s. They are called bits. And they can only understand instructions given in the binary format. Every set of instructions that we give to sleep you are translated into a set of instructions that tell the processor to do a particular job. You should understand that different types of processors have different types of instruction sets. For example, the Pentium processors will only understand their instructions set. It is a similar case with Macintosh. In the very beginning, programmers had to write their instructions in the binary language. It was very time-consuming and challenging. Luckily, we don't have to go through all of that.

Assembly language
As the machine language is tough to deal with, a new language called the assembly language was invented. Here every instruction is given a short name, and the variables are replaced by names and not by binary digits, making it easy for a programmer to write code. You may ask how you can understand and assembly language. It cannot. The assembly language will be converted into machine language with the help of an assembler. You should remember that each CPU has its assembly language. So, the assembly language of a CPU cannot be run on a different CPU. Even assembly language has got its drawbacks. They require large sets of instructions, even for simple tasks.

High-level language
The high-level programming languages came into existence to solve those problems that the assembly and machine languages were causing. These can run on any computer. High-level languages come with A program called a compiler. The role of a compiler is

to generate an executable file or a program that a CPU can directly understand. These programs are standalone programs. The modern compilers that are available today are very efficient and fast and converting the code into an executable format. Some of the programming languages use the interpreter. The job of an interpreter is to execute the code but without compiling it to the machine code. But here we won't learn about an interpreter. Any given programming language can be interpreted or compiled. Languages like C, C++, and PASCAL are compiled while scripting languages like JavaScript and Perl are interpreted.

Ask this question, and you will get a dozen different answers, but, put simply, C++ is a compiled, object-orientated computer-programming language. If you are new to programming, even that has probably left you somewhat confused, so let's break it down a bit more.

There are two types of programming language – one is interpreted, and the other is compiled. Interpreted languages are run on an interpreter – this reads the code and executes the commands inside one at a time. The CPU reads compiled languages on your computer instead of a program. Computers cannot read letters; they can only read binary numbers so, when you write your code, it has to be translated into binary first. This is called compiling.

The object-orientated part of the explanation just describes how the code is structured, but I will be going into more detail on that later on.

What does this all mean to me?

To be successful at learning C++, you need to use both a text editor and a compiler. If you are using a Windows PC to do this on, you can use Notepad, which is already installed, on your computer as your text editor and, as your compiler, which is going to convert your code into an executable format, you should use GNU C/C++.

Linux/Unix

A simple way to check whether GCC is installed on your Linux/Unix computer is to bring up the command line and type in:

$ g++ -v

If GCC is installed you will get the following message, or something very like it:

Using builtin specs

Target: i386-redhat-linux

Configured with: ../configure --prefix=/usr ....

Thread model: posix

gcc version 4.1.2 20080704 (Red Hat 4.1.2-46)

If GCC is not installed, you can follow the instructions on http://gcc.gnu.org/install/ to do so.

Mac OS X
The best way to get GCC for Mac OS X is to go to http://developer.apple.com/technologies/tools, download xCode, and follow the installation instructions.

Windows
To get GCC on your Windows PC, you must first download MinGW from http://www.mingw.org. Make sure you download the latest version of MinGW, as well as installing gcc-core, binutils, gcc-g++, and MinGW runtime – these are minimum requirements.

One more thing you could consider is an IDE – Integrated Developments Environment. This text editor features an integrated compiler and syntax highlighting. The best one to use is Code: Blocks.

Is there anything else I should know?
Yes. C++ is very rigid in its syntax, and everything is case sensitive, so be careful what you are typing. In C++, the word "Hello" means something different to "hello" but this is something you will learn more about later on in the book. For now, let's get straight on with the programming.

Before we go into too much detail, let's get you started in writing simple code. This is a typical example for all types of computer programming tutorials, so, without further ado, let's break it down. First of all, the code is written as such:

```
#include <iostream>
using namespace std;

int main ( )
{
        count << "Hello, World!";
        return 0;
}
```

Line 1 is telling your computer that you want to link iostream library to your program. The library is a repository where little bits of code are stored, along with variables and operators on occasion. The iostream library has code in it that lets us move characters to a text stream, which then appears on the screen.

Line 2 tells the computer that we want to use the standard namespace – this is where your code will be stored, useful if you write two bits of code with the same name, Line 4 is the start of the code. Everything that goes inside the braces is what the compiler reads and translates – this line must go in every single program you write; otherwise, nothing is going to happen.

The next stage is the printing of the program, which, in this case, is "Hello, World!" This is where we use some of the code stored in iostream – a piece of code called cout. The << that comes after the code is an operator, and this one translates to "output" which means we are telling the computer to send the code to a specified stream. The printing "Hello, World!" is called a string, which is nothing more than a character sequence enclosed in double quotations. Notice that the line ends with a semicolon – this is telling the computer that the code is finished, and if it is not used, the program cannot be compiled.

Line 8 is always placed at the end of main, and it is to inform the computer that the block is completed.

Now that you have the source code for the program, it has to be compiled using your compiler. More detail later, first, I want to start going over what makes up your code. The next chapter details all that you will need to learn to become a successful C++ programmer.

Structure of the C++ program

As we have already discussed, a computer program is nothing but a set of instructions in a sequence that tell the CPU what to do.

Statements and expressions

In C++, the statement is nothing but the smallest and an independent unit in the program. It is also the most common form of instruction in a programming language. In C++, statements are a way to convey your instructions to the compilers to carry out a specific task. We use ';' to terminate a statement. In C++, there are many kinds of statements, and some of the most common types of statements are given below.

int x;

x = 5;

std::cout << x;

We will discuss the functions of the above statements in the later chapters.

Functions

Functions are nothing but statements in a sequence, which are executed in sequential order. There is a function called the main function, and it is a must in every C++ program. When any given C++ program is run, the execution of the program starts within the main function. Functions have very specific jobs to perform. And they can perform only those jobs. For you to become a good C++ programmer, you need to know how to use functions efficiently. In the next chapter, you'll learn more about functions.

Libraries and the C++ standard library

Libraries are nothing but a collection of code, which is pre-compiled. They can be used within different programs. If you wish to extend the capabilities of your program, you can add libraries.

The core C++ language is minimalistic and straightforward, but it also comes with a built-in library called the C++ standard library. The C++ standard library provides the user with additional functionalities.

Syntax and syntax errors

We all know that in any language, sentences are constructed following the grammatical rules of that particular language. Take the example of the English language where you normally end of sentences in a period. All the rules, which govern the construction of sentences in a language, is called syntax. Similarly, C++ also has syntax. Not following the syntax will simply leave your code invalid. It is the responsibility of the compiler to see that the basic syntax of the C++ language is followed. If you violate the basic syntax, the compiler will compile the code and will complain about the syntax violation. That issue is called a syntax error. The C++ compilers only compile the program if there are no syntax errors found. It will display all the errors that are found and suggest for you to make them valid so that it could execute the code without any errors.

Compiling your first program

There are two important things that you should know before you can go to your first program.

Your programs will be written in the .cpp files, which will be added to your project later. Projects are also called as workspaces or solutions. The code files and the IDE settings will be saved to your project.

You should start a new project by creating one as there are many kinds of projects. Before starting, you need to specify the type of project you wish to work on.
What makes the C++ code?

A variable is something that stores a value. It's like a box, each box has a name and category on it, and the category is the type of item stored in it – the name is the specific variable. Each variable has a specific type, name, and size, and these govern exactly what can be stored inside of it.

Line 7 is called an assignment. This assigns a value to the variable, and the = is the operator – this is what does the assigning. The value after the = is called a binary operator.

Note that there are no quotation marks around the cout statement this time. This is because, rather than displaying the string, we are outputting the value. Compiling the program now would print "42" on the screen.

As well as having a type a variable can also have a prefix added to the declaration and these changes how the variable behaves. Some will change the size of the variable while others let you use or block you from using negative numbers, while others do a completely different job. Almost all of them will be talked about here.

Size Changers

Two specific prefixes change the byte size of the variable – short and long. The idea behind this is to allow the variable to store more vales and bigger numbers. How big depends on the architecture of the computer that you are using:

On a 32-bit computer int and long int are both 4 bytes
On a 64-bit computer int is 4 bytes and long int is 8 bytes

On both architectures, short int is 2 bytes

There is nothing that dictates the size of the modifier for those particular keywords, except:

Short must be equal to or smaller than a normal variable

The normal variable must, in turn, be equal to or smaller than a long.

The only variable that can take a prefix is int

Even more confusing, you can omit int and just use long or short, for example:

short int s = 0;

int i = 0;

long int l = 0;

short s2 = 0;

long l2 = 0;

Sign Changers

Two other prefixes can restrict the values that are stored by a type – signed and unsigned. A signed variable can store negative and positive values, while an unsigned variable can only take positives.

Other Prefixes

There are some other prefixes that you can use in C++, not so common, and most of which we are not going to use and will not cover in this book. There are two very useful ones, though – const and static. The former is constant, never changing their value, and the latter keeps the value of the variable the same but only in between function calls; that's a topic for later on.

At the Core of The Language: Understanding C++

Computers run virtually everything around us today; at the core of computer or your preferred cellphone brand, are programs. For our learning process, I assume you are aware of what a computer program is and structures behind it. At the core of all programs is a computing language. This question then comes up. What exactly is C++?

C++ (pronunciation see plus plus) is at its core a general all-purpose programming language. What makes the language different is the fact that it integrates crucial object-oriented coupled with generic programming in addition to providing low-level memory management.

The language and its design have a bias towards systems programming, for example, embedded systems, and operating system kernels just to point out a few.

The language is a favorite of many programmers due to its performance, flexibility, and efficiency. The language has effective uses in many contexts such as servers, e.g., Web search SQL and e-commerce, desktop applications, performance-critical applications such as space probe and telephone switches, and more than anything else, entertainment software, which is where your favorite video game falls.

The language is termed as a compiled language with its implementations available on many platforms across the globe and in use in many multinationals such as Intel, FSF, Microsoft, and LLVM just to point out a few.

If you search the ISO (International Organization for Standardization), you will learn that C++ is standardized. ISO approved and published the current version of C++ in September of 2011 as ISO/IEC 14882:2011, which is informally known as C++11. C++ first was standardized in 1998 as ISO/IEC 14882:1198, then amended to C++03, and then again in 2003 as ISO/IEC 14882:2003. The above information though useful holds no importance to you.

What is more important is the current standard C++11, which supersedes all the others because of the integration of new features and a bigger standard library.

Everything related to computers, and in fact, everything else in the world has an origin. C++ is the brainchild of Bjarne Stroustrup at bell labs as early as 1979 (C). Every great innovation has a cause or purpose. Bjarne wanted a more flexible and efficient language such as C but also a language that provided higher-level features of organizing programs.

Here is a kicker; remember when I said that learning C++ would prepare you to learn other languages? Here is why. C++ has influenced many languages, including Java C#, and all later versions of C developed after 1998.

Before closing this first chapter, I should perhaps mention that C is the precursor to C++, and therefore, working with C++ will demand some knowledge of the latter.

Getting Started-Working with C++ For the First Time

While studying and learning C++ is exciting, there are some prerequisites before we embark on writing code. Before you can undertake any programming, you need an Integrated Development Environment (IDE). An IDE is a compiler that translates your code into a bunch of ones and zeroes that are the language or code your machine can comprehend.

If you do not already have a compiler installed, you can download a bunch of free and paid compilers.

Installing an IDE

An IDE forms a very central part of your programming. Therefore, there is some importance in us looking at some of the more fundamental aspects of an IDE, especially installation. As I have mentioned, you can use a bunch of IDE's. Most developers struggle with the question of which one, which IDE best suits me. Here is the thing, though; you can install multiple IDE on your system, so there is no one correct or wrong IDE. The most common and integrated compilers are Code::Blocks and Microsoft's Visual C++ 2005 Express Edition. Therefore, I can recommend you install either one of them, but if you have a compiler that you deem friendlier, better, or superior to these two, do not feel restricted by my choice; indulge yourself.

If you are programming on Windows, i.e., a Windows machine, chances are, the best option for you is Microsoft's Visual C++ 2010 Express edition that is downloadable from Microsoft's website. Perhaps it is also well to point out that the file you download from the website is simply a downloader; the actual IDE download commences when you run the downloader.

However, if you are working in an environment that has limited internet connectivity, and would like the download file for later use, with a little help from Google or your favorite search engine, you can find an offline installer.

P.s. I do not recommend downloading files from third parties

It is also important to note that most of the functions of C and C++ we shall look at are contained in the documentation file that comes with the installation of MSDN express. However, if you do not install MSDN on your drive, Visual C++ will use an online version available here. Depending on your internet bandwidth, the files may take some time to download, and once the download and installer are done, Windows will prompt for a reboot, so reboot the system.

You will probably notice a lag in your computer response time, a sort of "hang"; you should not be alarmed, or force shutdown since the lag is because your system is installing all the resources required to run the environment. Also, once your system boots up and opens the Windows desktop, you may find it necessary to rerun the installer similarly to the installation bit we just completed.

However, this time, the installer will not download the files; it will simply install the program.

On the other hand, you may be developing on Windows but are writing programs for porting with the Linux kernel or vice versa; in this case, your best option is to use Code::Blocks, the open-source, cross-platform free IDE that will run on both Windows and Linux.

Nevertheless, there is a key point to note for Windows users. If you intend to write C++ code on Windows using Code::Blocks, you should make a point of downloading the MinGW bundled version. If Code::Blocks does not tickle your fancy, you can also use Bloodshed's Dev-C++, which is also cross-platform, i.e., runs on Windows and Linux.

So far, we have covered almost all the other platforms except Mac OSX. If you are an OSX user, you can use Xcode or Eclipse. C++ on Eclipse is not automatically set; it will require downloading and installing optional C++ components.

The IDE installation part of programming is probably the hardest part of any programming tutorial. Now that we have completed that part, you are ready to begin coding your first program.

Your First C++ Program

Welcome to the most exciting part, perhaps the part that made you pick up this book, writing programs using C++. Like most computer language tutorials, our first focus will be to create a hello world program.

While the "Hello world" code may sound like a cliché, in our program creation, it is the best illustration of "spinal code" of virtually any and every program you shall create way after I am through teaching you my bit. Regardless of the complexity or simplicity of your program or code, it must have a few basic lines of code.

```
#include<iostream>
using namespace std;
```

The above code is what we refer to as a pre-processor directive. If you look at the code, you will notice that there is an 'i' and 'o' 'iostream'; they stand for input and output, respectively, and are an absolute requirement in any program you create using the C++. On the other hand, 'namespace std' informs the compiler that you intend to use the standard library. If you fail to use the code line to specify this, you will limit yourself and not be able to use cin, cout, or endl, which we shall look at in a little while.

```
int main ()
{
```

On the other hand, the code above indicates where our main program begins, which is where most of the coding happens. If you are wondering, the int means integer. An integer is what will pass back to the program.

It is also important to note that you should not overlook the parenthesis (); they are important even if they are empty. In the same breath, you should also not overlook the curly brace {, because it is necessary. Now that we have understood that, using the console out, let us print something to the screen.

```
cout<<"Awesome! My first program in C++ "<<endl;
return 0;
}
```

I do not know if you are familiar with the C language, but the cout in the code above is the printf statement. The code (cout) prints to the screen anything, text, or otherwise, that is within the quotation's marks. On the other hand, endl indicates the end of the line (it is similar to pressing the enter button while working on a word document file). The semicolon is also required, so do not forget it. In the code above, return 0 acts as the integer that returns to the int main. Not to sound like a broken record, do not forget to end in with the curly bracket because they signify the end of the main.

In the above code, here is how our main program code would appear.

```
#include <iostream>
using namespace std;

int main()
{
cout<<"Awesome! My first program in C++"<<endl;
return 0;
}
```

This may come as a surprise to you but, if you have followed everything we have done so far, you have just created your first program.

SURPRISE! The code we just used and created is the basic structure of all programs; think of it like learning the alphabet in kindergarten. You should make every effort to understand and memorize it. Secondly, you should remember in which order to use cout and endl. Do not forget to declare using namespace std; otherwise, the compiler will return an error.

However, there is a workaround. If you do not use namespace to declare, then you must use **std::** to declare each cout, cin, and endl.

Here is an example of what I mean by this. For clarity, we shall highlight the difference in the program in blue.

#include <iostream>

int main()

std::cout<<"Awesome! My first program in C++"<<std::endl;
return 0;
}

If you are not very conversant with programming in general or C++, you will realize that cout, cin, and endl will come up a lot in your coding. Opting to use the method I have just described will be tedious. Therefore, you should opt to instruct the compiler by using the namespace std to make your work faster.

Chapter 2

Declaring constants

C++ Constants

Constants are fixed values that programs may not be allowed to alter, and they are referred to as literals. A constant can belong to any of the available basic data types. Note that are just regular variables with the difference being that the values of constants cannot be changed after definition.

To define a constant using the **#define preprocessor**, we use the syntax given below:

#define identifier value

The identifier is the name of the constant while the value denotes the value assigned to the constant. The following program describes how this can be done:

#include <iostream> **using namespace std**;

#define WIDTH 7

#define LENGTH 11.

int main() {

int rectangle_area;

rectangle_area = LENGTH * WIDTH; cout <<"The area of the rectangle is "<<rectangle_area;
return 0; }

In the above example, we have defined two constants, namely WIDTH, and LENGTH. We have assigned values of 7 and 11 to these, respectively. We have also defined the variable rectangle_area and assigned it the value of the product of the two constants. We have then printed out the value of this variable.

When executed, the code should return the following as the output:

The area of the rectangle is 77

Using the const Keyword
We can also define a constant by use of the const keyword. To do this, we have to use the syntax given below:

const data_type variable = value;

Note that we have to define the data type of the constant. The value of the constant this time has been assigned using the assignment (=) operator. Let us give an example demonstrating this:.

#include <iostream> using namespace std;

int main() {

const int LENGTH = 7; const int WIDTH = 11; int rectangle_area;

rectangle_area = LENGTH * WIDTH; cout <<"The area of the rectangle is "<< rectangle_area; return 0; }

We have just created the same example, but this time, the constants have been defined using the const keyword.

When executed, the code should return the following result:

The area of the rectangle is 77

Note that programmatically, it is always good to name constants in the upper case.

Working With C++ Data Types

Now that we have created your very first program, there is something else of equal importance, if not greater, than spending every waking hour writing code; documentation.

Every program needs documentation, if not to help other programmers understand your program, to help you understand your program, years after writing it. That may sound like revealing your magic "code writing skill" but in the programming world, you are never a success until other programmers pay you accolades. When using the C++ language, you can add documentation in the form of notes to your program by using two slashes "//"the compiler will ignore anything you write between these two slashes on the same line.

Throughout your coding life and work, I would recommend you to document your work as often as possible. Additionally, you will also notice that most of the notes you write will have a different color from the code. If your compiler does not do this automatically, you should do it manually to avoid confusion.

Here is another fun fact you are probably not aware of; almost every program uses variables. Too "mathematical" for you? A variable in programming is the same as a variable in math, a letter used to hold a value. In programming, you use variables to hold different data types that you must declare before you run the program.

Therefore, to fix this, you must assign x value in the following manner:

x=12;

If we use this example in our program, each time we use the variable x, it will be similar to using the number 12 unless we specify otherwise. With this, when you want to print something to the screen, here is how you do it.

Note: in this case, we are printing x to the screen.

cout<<x;

If you have prior experience with the C language, it will not pass you that this is different. For one, there is no printf, and secondly, there is no place holder. This should not be cause for alarm. In the code above, the 'c' in cout is a placeholder for "console" and therefore, cout means console out. Let us assume a scenario whereby you do not want to assign a variable to your program but would like the program to assign a variable value while it is running. While this may sound very mathematical and complicated, you can easily achieve this by using cin, which, as you may have guessed, stands for the console in.

int x;
cin>>x; //Ask the user to input a value for x

As a programmer, you have to be very keen. Look at the code above and tell me if you notice anything different from what we have learned so far. Did you notice anything? The arrows in the above code are pointing in the opposite direction from when you use cout. In the above, we did not assign a value to the integer called x. This is because the program will prompt the user for a number value that will then be stored as the value for variable x. Here is how our program should now look like.

#include <iostream>

using namespace std;

int main ()
{

```
int x
cout<<"Please enter an integer"<<endl;
cin>>x;//the user will enter a value that will be stored in 'x'
cout<<"Your number is ";
cout<<x;//the program will print the value of 'x' to the screen
return 0;
}
```

The program above first declares an integer, integer 'X', then it asks you to enter a value that once entered, displays on the screen. In an instance where you or someone else wants to manipulate the variable either through multiplication or through division, you, the programmer, would have to declare some more variables and coin some form of expression. Let us add some code to the program above.

```
#include <iostream>

using namespace std;

int main ()
{
int x,y,z
cout<<"Please enter an integer"<<endl;
cin>>x;
cout<<"Please enter an integer"<<endl;
cin>>y;
z=x/y;//the value of 'z' equals 'x' divided by 'y'
cout<<"Your number is ";
cout<<z;
return 0;
}
```

In the above, the program declares three integers and equals variable 'Z' to the division of 'X' and 'Y'. When the program user assigns a value to X and Y, the program then displays the division held in variable 'Z'. While the above code may work ok, there are flaws bound to arise from Z being equal to the division of X and y. In the above instance, if the user assigns 4 as X and 2 as Y, that would not present a problem because 4 is divisible by 2. However, what would happen if the values were 6 and 4? Leaving the program code in its current state would prompt a 0 answer mainly because so far and as of right now, X, Y, and Z only hold integer numbers without decimals. In the above example of the user assigning 6 and 4, it is evident that they are not divisible, and subdivision will place a decimal. So, how do we fix this? The answer to this is relatively simple; rather than declaring Z, which is an integer that cannot hold values with decimals, we can declare it as a double. I would be tempted to say that double data types are superior to integers in the sense that double data types are capable of holding decimal numbers. Here is how our program would look like.

```cpp
#include <iostream>

using namespace std;

int main ()
{
int x,y;
double z;
cout<<"Please enter an integer"<<endl;
cin>>x;
cout<<"Please enter an integer"<<endl;
cin>>y;
z=x/y;
cout<<"Your number is ";
cout<<z;
return 0;
}
```

While the program code looks so much better so far, there is also something not entirely correct with it, and if you run the program, you will still not get the right answer. Can you guess why? Because of something called truncation. This might sound complicated but is, in fact, not. In the above program code, X and Y are integers. As we have seen, you cannot get the correct answer if the two integers result in a decimal.

Simply put, this is the meaning of truncation. To fix this, one of the variables has to be double type. Doing this prompts the compiler to recognize this and convert the other integer into a double albeit temporarily. Alternatively, you can simply declare each variable as a double; this is something at your discretion. Here is how the full program code would look.

```cpp
#include <iostream>

using namespace std;

int main ()
{
int x;
double y,z;
cin>>x;
cout<<"Please enter an integer"<<endl;
cin>>y;
z=x/y;
cout<<"Your number is ";
cout<<z;
return 0;
}
```

Types of constants in programming
Integer Constants
First thing first, integers are stored in binary formation. You'll code integers as you use them in your daily routine; for example, you will code eight simply as 8.

The following table will show you different integers, their values in programming and their data types

Value in programming	Number	Data Type
98	+98	int
-865	-865	int
-68495L	-68495	long int
984325LU	984325	unsigned long int

Character Constants
Whenever, you'll find an integer, closed between two single apostrophes, this would be character constant. Moreover, there is a chance that you'll find a backslash "\" between those apostrophes.

For most machines, ASCII character set is used, i.e.

ASCII Characters	Symbolic Display
Null character	"\0"
newline	'\n'
horizontal tab	'\t'
alert (bell)	'\a'
backspace	'\b'
form feed	'\f'

vertical tab	'\v'
single-quote	'\''
backslash	'\\'
carriage return	'\r'

Float Constants

Float constants are stored as two parts in memory as float constants are numbers having decimal parts. The first part, they obtain in memory is significant, and the second is the exponent.

Float constant's default type is "double". You must write code to specify your desired data type, i.e., "float" or "long double". We may remember that "f" or "F" is used to represent float, and "l" or "L" is used to represent long double.

In the following table, some of the examples of float, double and long double are shown:

Value in Programming	Number	Data Type
.0	0.00	double
0.	0.00	double
3.0	3.0	double
5.6534	5.6534	double
-3.0f	-3.0	float
5.6534785674L	5.6534785674	long double

Boolean Constants

These constants are predefined keywords, and they cannot be defined or declared by the programmer. It has two predefined constants, "True" and "False". In programming, we represent this kind of constant as "bool".

Programming Constants

In this part, we are going to understand different programming constants and ways to write and define constants in a C++ program. Usually, there are three types of programming constants.

Defined Constants

A way to define a constant in a C++ program is to use a precompiler statement "define". Like every other precompiler directive, it starts with a "#". For example, a traditional precompiler directive for "define" would be: #define TABLE_SIZE 150.

Define directives are usually placed at the beginning of the program so that anyone reading your program can find them easily.

Memory Constants

Another way to code constants is by using a memory constant. These constants use a C++ type qualifier to remember that the specified data cannot be changed.

C++ programming provides us with an ability to define named constants. We just have to add type qualifier in our code, before constant. For example: **Code**.

```
#include <iostream>

using namespace std;

#define val 50

#define floatVal 9.7

#define charVal 'K'

int main()

{

    cout << "Integer Constant in our code: " << val << "\n"; cout << "Floating point Constant in our code: " << floatVal << "\n"; cout << "Character Constant in our code: "<< charVal << "\n"; return 0;

}
```

Output:

In the case of this code, a console screen will pop up with the output:

Integer Constant in our code: 50

Floating-point Constant in our code: 9.7

Character Constant in our code: K

Literal Constants

A literal constant is a constant which is unnamed and used to specify your desired data. As we know, a constant cannot be changed, so we just have to code its data value in a statement.

A literal constant is the most common form of constant. Here is a table to show a different kind of literal constants.

Values	Type
'C'	Character Literal
7	Numeric Literal 7
C + 8	Another Numeric Literal (8)
5.6534	Float Literal
Hello	String Literal

Chapter 3

Functions

You may remember from math class at some point where you had to deal with functions, like f(x) = 2x + 5. f(0) would have been 5 (2(0) = 5), f(1) would have been 7 (2(1) + 5), f(2) would have been 9 (2(2) + 5), and so on.

Functions in C++ work similarly.

Functions operate and then often return a value. They don't always have to return a value. However - functions that don't return a value are void functions.
First, take a look at this chunk of code. It doesn't have to make sense right now; we're going to go through it. Create a new project and erase the contents of main.cpp, and type this in.

```
#include <iostream>
using namespace std;

int getArea(int l, int w);

int main()

{

int length, width;

cout << "Enter the length, and then enter the width.\n";

cin >> length >> width;

cout << "The area is: " << getArea(length, width);

return 0;

}
```

C++

```
int getArea(int l, int w) {

    int area = l * w;

    return area;

}
```

So let's break this down line by line.

#include <iostream>
This includes the essential components for input and output.

using namespace std;

This denotes that we're using the standard input/output namespace.

int getArea(int l, int w);

This is where we declare our first function. C++ is procedurally compiled, so you can't just throw functions around willy-nilly. You have to declare them before you write them later, or they have to write entirely before your main function. For the sake of clarity, I prefer the initial declaration and then writing the function after my main function, but it's a personal choice. Regardless, we're creating a function here called "getArea". The int before the name denotes that it's going to return an integer value. Likewise, a float getArea(...) would return a float value, a char getChararacter(...) would return a character, and so on.

Within the declaration, we've given two variables, called arguments. Much like how in f(x) equations, x was the thing that the equation modified, the argument variables are what our C++ reads in and utilizes.

int main()

{

This is the start of our main function. In case you haven't figured it out, every program must have a main function. It's the entry point for your code, and the compiler will actively seek it out.
int length, width;

We declare two variables, length, and width;

cout << "Enter the length, and then enter the width.\n";

```
cin >> length >> width;
```

We ask the user to input the length and width, and then accept their input.

```
cout << "The area is: " << getArea(length, width);
```

This is where a bit of further explanation is needed. The output stream puts out a value we give it, right? Since functions simply return a value, we can put them straight in the output stream.

Also, since this function returns an integer value, you could create a new integer and then assign its value as the value of the function, like this: int area = getArea(length, width);

You could then just output.

```
cout << "The area is: " << area;
```

This is a waste, though, because this variable doesn't need to be created for this program. If you do, however, need to store a variable such as the area of a rectangle, you could do exactly that.
return 0;

}
Simple, every main function must return 0.

```
int getArea (int l, int w) {
```

This mirrors the function declaration earlier and is where we start writing it.

```
int area = l * w;

 return area;
```

We create a new variable called area, which takes the two arguments from the function header and multiplies them to get its value. We then return that same integer.

We also could have simply written this:

```
return l * w;
```

Both are valid ways of returning this function and will give you an appropriate result.

That's a cursory introduction to functions. They're a vital way of structuring your program. Programs, where everything is only in one function, are tedious to read and challenging to maintain. Programs, where things are more split up, are easier to understand and manage by far.

Chapter 4
Polymorphism

Polymorphism means to exist in many forms. In programing, polymorphism occurs when we have many classes that are related to each other through inheritance.

In C++, polymorphism means that a call to a function will lead to the execution of a different function depending on the type of object that has invoked the function. Consider the example given below in which we have a base class derived or inherited by two other classes:

```
#include <iostream> using namespace std;

class Figure {

protected: int breadth, length;
public: Figure( int x = 0, int y = 0) {

breadth = x; length = y; }

int area() {

cout << "Area of parent class is :" <<endl; return 0; }

}; class Rectangle: public Figure {

public: Rectangle( int x = 0, int y = 0):Figure(x, y) { }

int area () {

cout << "Area of Rectangle class is :" <<endl; return (breadth * length); }

};
```

C++

```cpp
class Square: public Figure {

public: Square( int x = 0, int y = 0):Figure(x, y) { }

int area () {

cout << "Area of Square class is :" <<endl; return (breadth * length); }

};

// Main function
int main() {

Figure *figure; Rectangle rec(10,7); Square sq(5,5);

// store address of Rectangle figure = &rec;

// call rectangle area.

figure->area();

// store the address of Square figure = &sq;

// call aquare area.

figure->area();

return 0; }
```

The code will return the following output:

```
Area of parent class is :
Area of parent class is :
```

The output is incorrect. The reason for this is that the call of area() function has been set only once by the compiler as the version that has been defined in the base class. |This is referred to as static resolution or static linkage of the function call. The function call is fixed before the execution of the program. Sometimes, this is referred to as early binding since our area has been set during the compilation of the program.

Let us make a modification to the code so that the area() function is declared within the Figure class. However, we will declare it with the virtual keyword as shown below:

```cpp
class Figure {
protected: int breadth, length;
public: Figure( int x = 0, int y = 0){
 breadth = x; length = y; }
virtual int area() {
 cout << "Area of parent class is :" <<endl; return 0; }
};
```

This means you should now have the following code:

```cpp
#include <iostream> using namespace std;
class Figure {
protected: int breadth, length;
public: Figure( int x = 0, int y = 0){
 breadth = x; length = y; }
virtual int area() {
 cout << "Area of parent class is :" <<endl; return 0; }
}; class Rectangle: public Figure {
public: Rectangle( int x = 0, int y = 0):Figure(x, y) { }
int area () {
 cout << "Area of Rectangle class is :" <<endl; return (breadth * length); }
};
class Square: public Figure {
public: Square( int x = 0, int y = 0):Figure(x, y) { }
int area () {
 cout << "Area of Square class is :" <<endl; return (breadth * length);
```

};

// Main function

int main() {

 Figure *figure; Rectangle rec(10,7); Square sq(5,5); // **store address of Rectangle** figure = &rec;

// call rectangle area.

figure->area();

// store the address of Square **figure = &sq**;

// call aquare area.

figure->area();

return 0; }

The code should now return the following result:

```
Area of Rectangle class is :
Area of Square class is :
```

The compiler has considered the elements of the pointer rather than its type. The objects of rec and sq classes have been stored in the *figure; their respective definition of the area() function has been called.

This shows that each of the derived classes has a different implementation of the function named area(). This is known as polymorphism. You have more than one class with the same function name and the same parameters, but the implementation is different.

Notice that we used the virtual keyword to make the function virtual. When a virtual function is defined in the base class, and another version in the derived class, this will act as a signal to the compiler that a static linkage to the function is not needed. We only need a selection of the function to be called at any point in the program based on the type of object for which the function is being called. This operation is known as late binding or dynamic linkage.

Sometimes, you may need including a virtual function in the base class for it to be redefined in a derived class to suit the class objects, but you have no meaningful definition to give to the function in the base class.

In such a case, our virtual function area() in the base class can be changed to the following:

class Figure {

protected: **int** breadth, length;

public: Figure(**int** x = 0, **int** y = 0) {

  breadth = x; length = y; }

virtual int area() = 0; };

Notice the use of the = 0; in the function. This tells the C++ compiler that the function doesn't have a body. Such a function is known as a pure virtual function.

Object Destruction and Polymorphis

Object destruction is tricky, especially when it comes to the object being destroyed through an interface. You could have code like this, for example:

class Drawable

{

public:

virtual void draw () = 0;

};

class MyDrawable : public Drawable

{

public:

virtual void draw ();

MyDrawable ();

~MyDrawable ();

private:

int *mydata;

};

```
MyDrawable::MyDrawable ()

{

mydata = new int;

}

MyDrawable::~MyDrawable ()

{

delete mydata;

}

void deleteDrawable (Drawable *drawable)

{

delete drawable;

}

int main ()

{

deleteDrawable( new MyDrawable() );
}
```

What's going on inside deleteDrawable? Remember, when delete gets used, the destructor is called, so the line that reads

delete drawable;

makes a function call on the object. But how does the compiler know where the MyDrawable destructor is? The compiler doesn't know what the exact type of the drawable variable is, but it does know that it's a Drawable, something that has a draw() method. All it knows is how the MyDrawable destructor itself can found the destructor associated Drawable. As MyDrawable allocates memory in its constructor, to free the memory, the MyDrawable destructor must run.

Okay, so you could be thinking that this is what virtual functions are designed to fix. That's correct. What we should be doing is declaring the destructor virtual in Drawable; that way the compiler will know that it has to look for a destructor that has been overridden when delete has been called on a pointer to a Drawable: class Drawable.

```
{

public:

virtual void draw ();

virtual ~Drawable ();

};
class MyDrawable : public Drawable

{

public:

virtual void draw ();

MyDrawable ();

virtual ~MyDrawable ();

private:

int *mydata;

};
```

When we make the structure in the superclass virtual, and when delete frees up a Drawable interface, the overridden destructor gets called.

As a rule, when you make any superclass method virtual, the superclass destructor should be made virtual as well. Once one method has been made virtual, you are saying that the class can be passed around to methods that will take an interface. The methods can then do whatever they want, and that includes deleting the object, so making the destructor virtual makes sure that the object is cleaned up properly.

Chapter 5

Operator type and overloading

Operators in C+

C++ provides a rich set of operators that perform a number for different functions on constants and variables.

C++ has mainly following types of operators.

Arithmetic Operators as addition, multiplication, etc.

Relational Operators as less than, greater than, etc.

Logical Operators as AND, OR, etc.

Bitwise Operators

Assignment Operators
Arithmetic Operators

C++ mainly provides 7 arithmetic operators. We will discuss each operator with two variables x and y where x has value as 45 and y has value has 25.

Arithmetic Operator	Meaning	Example
+	Simple mathematical addition of operands	x+y gives 70
-	Simple mathematical subtraction of operands	x-y gives 20
*	Mathematical multiplication of operands	x*y gives 1125

/	Mathematical division of operands	x/y gives 1
%	It is a Modulus Operator which provides remainder after integer division	x%y gives 20
++	Named as Increment operator and increases the value of operand by 1	x++ gives 46
--	Named as Decrement operator and decreases the value of operand by 1	x - - gives 44

Relational Operators

C++ mainly provides 7 Relational operators. These operators provide a resultant value as TRUE or FALSE and generally used in condition checks. We will understand the meaning of each operator with two variables x and y where x has value as 45 and y has value has 25.

Relational Operator	Meaning	Example
==	Equality operator used to check whether two operands are equal or not. Will result as TRUE if they are equal otherwise false	(x == y) is FALSE.
!=	Inequality operator used to check whether two operands are unequal or not. Will result as TRUE if they are unequal otherwise false	(x != y) is TRUE.
>	Greater than operator and check if left operands is greater than right or not. If left is greater, returns TRUE otherwise FALSE	(x > y) is TRUE.
<	Less than operator and check if left operands is lesser than right or not. If left is lesser, returns TRUE oth-erwise FALSE	(x < y) is FALSE.
>=	Greater than or equal to operator and check if left operands is greater than or equal to right operand or not. If left is greater or equal , returns TRUE otherwise FALSE	(x >= y) is TRUE.
<=	Less than or equal to operator and check if left operands is lesser than or equal to right operand or not. If left is lesser or equal, returns TRUE otherwise FALSE	(x <= y) is FALSE

Logical Operators

C++ supports the following 3 logical operators, which we will explain using two variables x and y where x having value as 0 and y having value as 1.

Logical Operator	Meaning	Example
||	Logical OR Operator. Returns TRUE if at least one operand is non-zero	(x || y) returns TRUE
&&	Logical AND Operator. Returns TRUE if both operands are non-zero	(x && y) returns TRUE
!	Logical NOT Operator and reverse the logical value for its operand. If operand is zero makes that one as non-zero and operand is non-zero makes that as zero.	!(x && y) returns TRUE

Bitwise Operators

Bitwise operators perform their functionality bit by bit. Bitwise AND operator performs AND operator on every two corresponding bits of two operands and Bitwise OR performs OR operator on every two corresponding bits of two operands

x	y	x | y	x & y	x ^ y
1	1	1	1	0
0	0	0	0	0
1	0	1	0	1
0	1	1	0	1

Assume if x = 13; and y = 60;

x = 0000 1101

y = 0011 1100

x|y = 0011 1101

x&y = 0000 1100

Table showing bitwise operator with p= 60 and q = 13

Bitwise Operator	Meaning	Example
|	Bitwise OR Operator, gives 1 if any bit 1	(p | q) gives 0011 1101 which is 61
&	Bitwise AND, gives 1 if both bits are 1	(p & q) gives 0000 1100 which is 12
^	Bitwise XOR Operator, gives 1 on-ly when exactly one bit is 1	(p ^ q) gives 0011 0001 which is 49
~	Bitwise Ones Complement, change the logical value for bit.	(~p) gives 1100 0011 which repre-sents -61
>>	Bitwise Right Shift Operator which moves the left operand's value to-wards right by the bits specified which is a right operand	p >> 2 gives 0000 1111 which rep-resents 15
<<	Bitwise Left Shift Operator which moves the left operand's value to-wards left by the bits specified which is a right operand.	p << 2 gives 1111 0000 which rep-resents 240

Assignment Operators

Assignment Operator	Meaning	Example
=	Known as assignment operator which assigns the value of right hand side expression to left oper-and	p = q + r assigns value for q+r in-to p
-=	Involves first in subtraction then assignment	p -= q is similar to p = p − q
+=	Involves first in addition then as-signment	P+= q is same as p = p + q
/=	Firstly division is performed and after that assignment	p /= q is same as C = C / A
*=	Firstly multiplication is performed on the given operands and after that assignment	p *= q is same as p = p * q
%=	Firstly modulus is performed on the given operands and after that assignment	p %= q is same as p = p % q

>>=	Firstly Right Shift Operation is performed and after that assign-ment	p >>= 3 is same as p= p >> 3
<<=	Firstly Left Shift Operation is performed and after that assignment	p <<=3 is same as p = p<< 3
&=	Firstly Bitwise AND is performed on the given operands and after that assignment	p &= 3 is same as p = p & 3
^=	Firstly Bitwise Exclusive OR is performed on the given operands and after that assignment	p ^= 3 is same as p = p ^ 3
|=	Firstly Bitwise Inclusive OR is performed on the given operands and after that assignment	p |= 3 is same as p = p | 3

Operators Precedence and Associativity

Every operator has precedence and associativity associated with it. Precedence of operators is used in determining the sequence of operators for evaluating an expression in case expression has several operators. In an expression, the operators having higher precedence are evaluated first.

Associativity is the concept for determining the evaluation order of operators when there are multiple operators with the same precedence, then associative determines either to be evaluated from left to right or right to left.

Category	Operator	Associativity
Postfix Operator	[] () -> - - ++ .	Left to right
Unary Operator	+ ! ~ - - ++ (type)* sizeof &	Right to left
Multiplicative **Operator**	* % /	Left to right
Additive **Operator**	- +	Left to right
Shift Operator	>> <<	Left to right
Relational **Operator**	< > <= >=	Left to right
Equality **Operator**	!= ==	Left to right
Bitwise AND **Operator**	&	Left to right
Bitwise XOR **Operator**	^	Left to right
Bitwise OR **Operator**	|	Left to right
Logical AND **Operator**	&&	Left to right
Logical OR **Operator**	||	Left to right
Conditional **Operator**	?:	Right to left

Assignment Operator	+= = -= /= *= %= <<= >>= &= |= ^=	Right to left
Comma Operator	,	Left to right

Overloading

In C++, we can specify more than one definition for a function or operator within the same scope. These are processes known as function overloading and operator overloading, respectively.

An overloaded declaration is done within the same scope as the previous declaration, but both definitions have different arguments and a different implementation/definition.

After calling an overloaded function or operator, the compiler will determine the most appropriate definition to use by comparing the types of arguments that you have used to call the function or operator with the types of parameters that have been specified in the definition. The process of determining the most appropriate definition to use is referred to as overload resolution.

Operators Overloading

C++ allows you to redefine or overload the majority of its in-built operators. This means that programmers can use operators with user-defined types.

Overloaded operators are simple functions with special names, that is, the operator keyword and then symbol of the operator that is under the definition. Just like a function definition, an overloaded operator will have a return type and a list of parameters. The definition of an overloaded operator is demonstrated below:

Rectangle operator+(**const** Rectangle&);

In the example given below, we are demonstrating the process of operator overloading by the use of a member function. We will pass an object as an argument, and the properties of the object will be accessed using the object. The object that will call the operator will be accessible by use of this operator as demonstrated below:

#include <iostream> using namespace std;

class Rectangle {

public: **double** getArea(**void**) {

return length * breadth; }

void setLength(**double** len) {

length = len; }

void setBreadth(**double** bre) {

breadth = bre; }

// Overload the + operator to add two Rectangle objects.

Rectangle operator+(const Rectangle& r) {

Rectangle rect; rect.length = this->length + r.length; rect.breadth = this->breadth + r.breadth; return rect; }

private: **double** length; // Length of a rectangle **double** breadth; // Breadth of a rectangle };

// Main function

int main() {

Rectangle Rectangle1; // Declare Rectangle1 of type Rectangle **Rectangle Rectangle2**; // Declare Rectangle2 of type Rectangle **Rectangle Rectangle3**; // Declare Rectangle3 of type Rectangle **double area** = 0.0; // Store the area of a Rectangle here

// Rectangle 1 specification **Rectangle1.setLength**(6.0); **Rectangle1.setBreadth**(7.0);

// Rectangle 2 specification **Rectangle2.setLength**(12.0); **Rectangle2.setBreadth**(13.0);
// area of Rectangle 1

area = Rectangle1.getArea(); cout << "Area of Rectangle1 : " << area <<endl;

// area of Rectangle 2

area = Rectangle2.getArea(); cout << "Area of Rectangle2 : " << area <<endl;

// **Adding two objects**: Rectangle3 = Rectangle1 + Rectangle2;

// area of rectangle 3

area = Rectangle3.getArea(); cout << "Area of Rectangle3 : " << area <<endl;

return 0; }

When executed, the code will return the following:

```
Area of Rectangle1 : 42
Area of Rectangle2 : 156
Area of Rectangle3 : 360
```

Consider the following section extracted from the above code:

Rectangle operator+(const Rectangle& r) {

  Rectangle rect; rect.length = this->length + r.length; rect.breadth = this->breadth + r.breadth; return rect; }

This is where the mechanism of operator overloading has been implemented. We have overloaded the + operator so that it adds the measurements of two rectangle objects. Here is another line extracted from the code:

Rectangle3 = Rectangle1 + Rectangle2;

In the above line, Rectangle1 will be added to Rectangle2. This means that the length of Rectangle1 will be added to the length of Rectangle2 to get the length of Rectangle3 while the breadth of Rectangle1 will be added to the breadth of Rectangle2 to get the breadth of Rectangle3. These two will then be multiplied to give the area of Rectangle3.

The meaning of + operator has been overloaded!

Chapter 6

Macros and template

Introduction to Templates
Templates are an important feature in C++ programming language, as it allows functions and classes to use generic types. They play an essential role in how functions and classes work by allowing them to define different data types without rewriting each other.

Using templates, you can write a generic code that can be used with any data type. All you need is to pass the data types as parameters. Templates promote code reusability and improve the flexibility of the program. You can create a simple class or a function and pass data types as parameters and implement the code to be used by any data type. Whenever you want to use the function, you make a call to the function and specify the return type.

For example, if you use **sort()** function to sort data in a warehouse, you can pass different data types as parameters to the function. You don't have to write and maintain multiple file codes to sort data. Based on the data type passed to the sorting algorithm, your data in the warehouse is sorted irrespective of that particular data type. Therefore, templates are used in situations where you need a code to be reusable with more than one data type.

Templates are of great importance when dealing with multiple inheritance and operator overloading. It is widely applied where code reusability is of prime importance.

Types of Templates
Templates are implemented in two ways:

- Using function templates

- Using class templates

A template can be defined as a macro. That is when you define an object of a specific type, the template definition of that type of class substitutes it with the appropriate data type. Templates are often referred to as parameterized class or functions since a specified data type will replace the defined parameter during execution time.

Class Templates
Class templates offer the specifications for creating a class based on the parameters. A class template accepts members with parameter type. That is, you can instantiate a class template by passing a given set of data type to its parameter list. Class templates are mostly used in implementing containers.

Syntax
Template <class identifier>

Class class-name

{

...

//class member specifications

...

};

Example:

Template <class M>
Class vector
{
  M*v,//type M vector
  int size, sum;

  public:
  vector (int n)
  {
  v=new M [size = n];
  for(int j=0; j<size; j++) v[j]= 0;
  }

  vector (M*a)
  {
  for(int j=0; j<size; j++)
  v[j]= a[j];
  }

  M operator* (vector & y)
  {

```
    M sum=0;
    for(int j=0; j<size; j++)
    sum+=->v[j]*y-v[j];
    return sum;
    }
};
```

A class template definition is similar to normal class definition except that in templates, you add a prefix template <**class M**> where M is the type. The template prefix tells the compiler that a class template is declared, which uses M as the type name in the declaration. In this case, the class vector is a parameterized class that uses type M as its arguments. Type M can be replaced by any data type or by using a user-defined data type.

If a class is derived from a class template, then it's called a **template class**. The objects of the template class are defined as follows: Classname<type> objectname (arglist);

Creating a derived class from class template is known as **instantiation process**. Before creating a template class, you have to debug the class template before converting it to a template class. The compiler checks for errors in a template class after instantiation has taken place.

Example: class template

template <Class TT>

Class Rectangle

{

TT width;

TT height;

public:

void setvalues(TT num1,TT num2)

TT area()

{
TT A;

A=width*height;

return A.

}

};

void Rectangle <TT> ::setvalues (TT num1,TT num2)

{

width=num1;

height=num2;

}

int main()

{

Rectangle obj1;

Rectangle <int> obj1(5,6);

Rectangle obj2;

Rectangle <float> obj2(5.6,4);
cout<< obj1.setvalues()

cout<<obj2.setvalues()

return 0;

}

//output

obj1=30

obj2=22.4

The above program creates a class of type TT. The <TT> in the (void Rectangle <TT> ::setvalues (TT num1,TT num2)) statement specifies that the function parameter is also a class template parameter and should always be included when calling a class.

Class Templates with Multiple Parameters

You can also have more than one generic data type in a class. You can declare the generic type class with a comma-separated list inside the template specification.

Syntax

Template <Class xy, Class xyz, ...>

Class class-name
{

  //function body;

};

Example: class with two generic types

```
template<class f1,class f2>

class numbers

{

 f1 j; f2 k;

 public:

 numbers (f1 x, f2 y)

 { j=x; k=y;

 }

 void print()

 {

 cout<<j<<"and"<<k<<"\n";

 }

};

int main()

{

 Numbers <float, int> numbers1 (2.53, 253);
```

Numbers <int, char> numbers2 (50,'Z');

numbers1.print();

numbers2.print();

return 0;

};

// output

1.53 and 253

50 and Z

Function Templates

Function templates work like a normal function except that the normal function works with only a single data type, while templates accommodate multiple data types. With templates, you can overload a normal function to work with different data types.

This makes function templates more useful since you only have to write a single program that works on all data types.

Templates are always expanded during compile time.

Just like a class template, you can create a function template with different argument types.

Syntax

Template <class type>

Return-type function-name (arguments)

{

Function body;

};

When defining function templates, you must include the template type in both the function body and parameter list when necessary.

Example1: Function to swap values

Template <Class T1>

Void swap (T1&num1, T1&num2)
{T1 val= num1; num1=num2; num2=val;};

Calling a template function works the same way as the normal function call.

Example 2: Implementation of function template

template <class test>

test max (test x, test y);

{

test results;

results=x>y? x:y;

return results;

};

int main()

{

int a=5, b=6, c;

double l=4, m=6, n;

c= max <int> (a,b);
n=max <double> (l,m);

cout<< c<<endl;

cout<<n<<endl;

return 0;

}

In this example, the test is the template parameter. The function max is called twice with different argument types (int and double). When the compiler instantiates, it calls the function each time by its type.

The output object produced (after instantiation of the template with specific type) will be of the same type with the parameters x and y.

The above program will sort values based on which is higher. If using an array of numbers, the sorting algorithm will be applied to sort the numbers from the smallest to highest.

Function Template with Multiple Parameters
Just like in class templates, you can have more than one generic data type separated by commas.

```
template<class T1, class T2,....>
returntype functionname(arguments of types T1,T2,...)
{
.......
.......
.......
}
```

Example: Temple class with multiple parameters

```
template<class Temp1,class Temp2>

void display( Temp1 j, Temp2 k)

{

cout<<j<<" "<<k<<"\n";

}

int main()

{

j=display1(2019, "EDGE");

K=display2(18.54, "1854);

 return 0;

}
// output
```

18.54 1854

Overloading Template Functions

You can overload a template function using its template function or using the ordinary function from its name. Overloading can be done through:

Calling an ordinary function that matches the template function.

Calling a template function that is built with the exact function.

Use normal overloading function on an ordinary function and make a function call to the one that matches the template function.

If there is no match found, an error message is generated by the system.

Example: Overloaded template using an explicit function

```
template <class M>

 void display(M n)

{

cout<<"Display template function:" << n<< "\n";

}

void display ( int n)

{

cout<<"Explicit template display: "<< n<<"\n";

}

int main()

{

 display (80); display (15.20);

 return 0;

}
```

//output

Explicit template display: 80 Display template function: 15.20

The function call display (80) calls the ordinary version of a display () function and not its template version.

Member Function Templates

When creating class templates, all member functions can be defined outside the class since the member functions in a template class are parameterized using type argument. Therefore, the function template should define the member functions.

Member functions (whether inline or non-inline) declared inside a class template are implicitly a function template. If a template class is declared, it inherits all the template functions defined in a class template.

Member function templates are defined in three ways:

Explicit definition in a file scope for each return-type used to instantiate a template class.

During file scope within the template parameters.

Inclined within the class.

A member function template can instantiate functions not explicitly generated. If a class has both a member function and an explicit function, then the explicit definition is given more priority.

Syntax

Template<class T>
returntype classname<T> :: functionname(arglist)
{
....
....
....
}

Example: Class vector with a member function template

Template<class T>
class vector
{ T*v;

```
int size; public: vector(int m); vector (T*a);
T operator*(vector & y);
}
//member function templates template<<classT>
vector<T> :: vector (int m);
{
v=new T[size=m]; for(int i=0; i<size : i++) v[i]=0;
}


template<class T>
vector <T>::vector(t*a)
{
for(int i=0; i<size ; i++)
v[i]=a [i];
}
template< class T >
T vector < T > :: operator*(vector & y)
{
T sum=0;
for (int i=0; i<size ; i++)
sum += this -> v [i]+y.v[i];
return sum;
}
```

Non-type Template Arguments

Templates can have single or multiple type arguments. You can also use a non-type argument template in addition to the type T argument. You can create a template argument using strings, constants, built-in types, and function names.

```
Template<class T, int size>Class array
{
T a [size];                    //automatic array initialization
//..........
//..........
};
```

The above template passes the size of an array as an argument. The compiler will only know the size of the array during the execution time. The arguments are specified during the creation of template class.

Template Classes

While template classes tend to be limited to library writers who want classes such as map and vector, everyday programmers can also benefit from being able to make their code more generic. You shouldn't use a template just because you can; instead, you should be looking for opportunities to get rid of classes that only differ by the type.

Yes, you are likely to find yourself writing more template methods than classes, but you still need to have some idea of how to use them, especially if you have a data structure of your own that you want to implement.

Declaring a template class isn't very different from a template function. Let's say that we want to build a small class that will wrap an array:

template <typename T> class ArrayWrapper

{

private:

T *pmem;

};

As with the template function, we begin with the declaration that a template is being introduced. We use the template keyword, and then the template parameters are added. We only have one parameter here, T.

Type T can be used whenever we want the type that would be specified by a user, just like with the template functions. When a function is defined for a template class, the template syntax also has to be used. So let's add a constructor called ArrayWrapper to our class:
template <typename T> class ArrayWrapper

{

public:

ArrayWrapper (int size);

private:

T *pmem;

};

// so that the constructor can be defined outside the class, we start by

// indicating that the function is a template

template <typename T>

ArrayWrapper<T>::ArrayWrapper (int size)

: pmem(new T[size])

{ }

We start with the basic template introduction and redeclare the parameter for the template. The difference is, the class name now has the ArrayWrapper template in it, which tells everyone that this belongs to a template class and isn't a template function on a standard class called ArrayWrapper.

In this implementation, the template parameter can be used to stand in for the provided type, just as with the template functions. However, the function caller will never need to provide the parameter—it gets taken from the first template type declaration. For example, instead of writing vec.size<int>()

or

vec<int>.size();

to get the size of a vector of integers, all you need to write is

vec.size().

Templates and Header Files

Up to now, the templates we have looked at have been directly written into .cpp files. What if we wanted to place our template declaration into the header file? Could we? The problem here is that code that makes use of template classes or template functions needs to be able to access the whole template definition for every single function call to the template as well as to every member function that is called on template classes.

This is not the same way that a standard function works; these only require the caller to know about the declaration. If, for example, you placed the Calc class inside its header file, you would also need to add the full constructor definition along with the add() method instead of putting them all into one .cpp file as you would normally. If you didn't do this, you would not be able to use Calc.

This is a rather unfortunate thing about templates, and it's all about how templates have been compiled. Most of the time, when the compiler parses them the first time, it pretty much ignores templates. It's only when a template is used that has a concrete type attached that the compiler generates the code for that type. To do that, the template has to be available to the compiler so the code can be generated. A result of this is that all template code has to be included in all of the files that will use that template. And, when a file that has a template is compiled, you may not learn of any syntax errors in that template until the template is used for the first time.

The easiest thing to do, when creating template classes, is to place all template definitions into the header file. It's also helpful if you use an extension other than .h, just to indicate that the file is a template—.hxx, for example.

Analyzing Template Error Messages

If there is one significant downside to using templates, it's that many compilers will throw out error messages that are not easy to understand when a template is misused, whether you write it or not. You could end up with pages of error messages just for one mistake.

Template error messages are not easy to read simply because the template parameters are expanded to their fullest, even those that you don't normally use, such as the default parameters.

There is a similar constraint on the second parameter for the vector template— this one requires a type that has support for more functionality than the humble basic integer gives us. The errors are all complaining about the multitude of ways that int just isn't a valid type for the parameter.

Finding the code line where the error is will always be the first step to working out where it all went wrong. More often than not, and even more so as you gain experience and confidence, a simple look at your code will tell you what the error is. If it isn't that obvious, you just keep going through the instantiation list until the proper error message appears: "error: int is not a class, struct or union type." The compiler expected a structure or a class, not any of the built-in types such as int. A vector should hold any type, so the error lies in the parameter given to the vector.

You should make sure you know how a vector is declared and that it only needs one template parameter.

Now that the problem has been found, we can fix it and then recompile the code. Sometimes you will be able to work out and fix a few errors at the same time, but where templates are concerned, the first error is usually the root of the problem for all others. Fix things one at a time, and you will find that any other errors fix themselves.

In our final example above, all those error messages resulted from one simple mistake: giving a second template parameter of an int.

Chapter 7

Classes

Introduction to Classes

At the heart of all object-oriented programming lies the notion of classes, objects that you create, and utilize throughout your code. This helps with modularity, with ease of use, with efficiency, and with higher-level programming that is far more logical and coherent.

In the early iterations of C, there was a primitive sort of class called a struct. Structs still exist in C++ but are mainly deprecated and shirked in favor of using classes instead. Nevertheless, structs are important to cover. C++ structs are also different from C structs. C structs didn't have much of the functionality which C++ structs do.

Before we jump into that, let's talk briefly about the notion of access modifiers.
We've talked about things like int, string, bool, and float called types. We created what could be considered somewhat of a type using enumerators. However, you can create an entire type that is made up of smaller pieces of data and innate functions. This is called an object.

Access modifiers describe what can change what within your code. There are three different access modifiers.

Public access means that the object and/or its internal data and functions can be accessed and modified by any other code in the program.

Protected means that the object and/or its internal data and functions can be access and modified by only its derivatives. (This will make sense when we get into inheritance and polymorphism.)

Private means that only the object itself can access and modify its data.

This is important for purposes of security, clarity, and code safety.

Structs and classes are both ways to create new objects. The primary difference between structs is that structs are public by default, and classes are private by default. Other than that, the differences are very negligible.

However, in the tech industry, structs have a fair bit of a negative connotation. Developers tend to see them as unprotected objects with limited functionality, while they tend to see classes as very functional objects with child classes that are well made and well structured. As such, it's generally better to create classes unless your object has very little in the way of functionality and simply contains very little data and innate function.

Look at this struct and declaration.

struct animal {

enum diet { herbivore, omnivore, carnivore };

int legs;

string name;

diet naturalDiet;

};

int main() {

animal dog;

dog.legs = 4;

dog.name = "Dog";

dog.naturalDiet = dog.carnivore;

}

This is the simplest way to declare a new object. You should not declare them within an existent function.

Let's go ahead and replace that struct with a class because we're going to be using classes going forward as they're generally a safer and better option than structs are.

Create a new project in CodeBlocks called "AnimalSimulator". Not the most creative title, but it'll work for what we've got to do. Redact the contents of main.cpp and import iostream and create your main method.

Above the main method, let's declare a class called Animal. You would do that very similar to how you'd declare a struct. For the sake of illustration, make it look like this:

class animal {

  string name;

};

Then within your main method, declare an instance and set its instance of name to the name of your favorite animal. Afterward, try to build.
You should have gotten an error that said "error: string animal::name is private".
Perfect. This is supposed to happen because now we have to fix this. This is where that whole notion of access description comes in.

Modify your class so that it looks like this:
class animal {

  public:

  string name;

}

Then try to build. It should work fine this time. That's because we made the string "name" public, which means other methods/functions outside of the class itself can access it.

This is generally considered bad practice though, so we're going to rewrite the class like so:

class animal {

  private:

  string name;

}

The best-practice way to modify values within a class is to use get/set methods. Here's the way the class would look if you implemented get/set methods.

class animal {

  private:

  string name;

  public:

```cpp
string getName() {

    return this->name;

}

void setName(string name) {

    this->name = name;

}

}
```

Now, in your main method, after declaring whatever animal it is, instead of directly modifying its name, you should instead use the get/set method like so (assuming your instance animal is called "dog"):
dog.setName("dog");

Then you can test this by printing out the name in your main method:

cout << "The name of my favorite animal is: " << dog.getName() << "\n";
The way these works is by returning or setting the value of the variable via a method from within the class. "this->" is called the this pointer, and it refers to the variable of a given instance of a class. Every object has access to its variables and can modify them directly via the 'this pointer'.

Anyway, your code should work stunningly. But what if you don't want to go through get/set for every instance of every class? Well, you use what's called an initializer function.

Modify your class so it looks like this:

```cpp
class animal {

    private:

    string name;

    public:

    animal(string name) {

        this->name = name;

    }
```

```
string getName() {

return this->name;

}

void setName(string name) {
this->name = name;

}

}
```

Now, in your main function, take out the chunk of code that says animal x followed by the x.setName("name") function. When you create an initializer function, you can set certain variables from the get-go.

For example, since your initializer function takes the argument of name, you can declare that when you declare the variable and circumvent the whole setName operation. Look at this code for an example:

animal dog("dog");

animal elephant("elephant");

animal bird("bird");

These are all separate instances of the class animal, and their names have been set without the use of a set function thanks to the function initializer.
This can take as many arguments as you'd like it to. This makes it incredibly easy to create new objects and pretty streamlined of a process too.
Now that we've spoken for a moment about declaring objects and things of that nature let's talk more about the specifics of what you can do with them. The object animal here, represents, of course, animals. We can give the entire class a set of methods that they can perform.

Let's think for a second about what every animal does. Every animal sleeps, eats, and drinks, right? So it wouldn't be very outlandish to include these within our class so that every member object of the class animal can perform these methods.

Let's say we had two specific kinds of food: meat and plants, both of which also represented by their respective objects.

We could create two different functions for this:

void eat(plant p) {

```
// code here

}

void eat(meat m) {

// code here

}
```

Even though these share a name, you can call either depending upon the type of object which you put include in the function call, and both will perform their respective code in response to the argument which you included. This is called function overloading. It's an essential technique in object-oriented programming that will help you to create functional and clear sets of code that are easy for people to use and understand. In the next chapter, we're going to start exploring some of the deeper things that you can do with classes, including the inherent qualities which make them both incredibly useful and incredibly practical in object-oriented programming.

Because it's unwieldy to create a large number of types while we're learning about classes in the first place, let's just leave "eat" as a void function that doesn't take arguments. Your code should look like this:

```
class animal {

private:

string name;

public:

animal(string name) {

this->name = name;

}

string getName() {

return this->name;
}

void setName(string name) {
this->name = name;
```

}

void eat() {

// eat food

}

}

In the next chapter, we're going to cover far more in-depth concepts regarding classes.

Objects and Classes

Classes are the special feature of C++, which casts C++ into Object-Oriented programming. Remember, C++ is not a pure object-oriented language because, in a pure object-oriented, we can perform no functionality without classes, but in C++, it is possible.

Class

It is a user-defined data type having data members and member functions that can only be accessed after an instance is created for that class. In general terms, a class is a blueprint that represents the states and behaviors for class's instances known as objects.

Data members are defined in class definition as variables and member functions as functions.

For example, a Class of humans, all humans can walk, run, eat, and they all have their specific color, weight, age. Here, walk, eat, and run are behaviors for humans and color, age, weight is their characteristics. There are several different human beings in this class, all having different names, but all of them have these characteristics and behavior.

So, the class is a blueprint that declares characteristics as data members and behaviors as member functions, and all the instances known as objects share all these characteristics and behavior.

Important points about classes

A Class name is conventionally started with an uppercase letter.

Class Human

A-Class has data members as variables and member functions as simple functions, and the access scope of these variables and data members is dependent on the access specifiers as Private, Public, or Protected, which we will discuss later.

We can define the member function of a class inside the definition of that class or outside the definition of class.

C++

Class in C++ is similar to structures in C with the only difference that by default, access control for classes is Private while the structure has public.

C++ provides all important features of object-oriented programming as Encapsulation, Abstraction, and Inheritance, etc. with classes.

Each object of a class has its separate copy of data members, and we can create as many numbers of objects for a class as needed.

Objects

Class is just a blueprint. Just with defining a class, no memory is allocated to it. Memory is allocated only after objects are created for that class. These Objects are instances having the data variables declared in that class and the member functions of that classwork on these instances.

Objects are initialized with special Constructors – special methods which we will study later. Destructors are other special functions that destroy the object when it goes out of scope.

class Mobiles {

int Price;

void Calling (){}

Void Messaging (){}

};

void main(){

Mobiles m1; // an object created for class mobiles }

Constructors

Constructors are special functions similar to class member functions and having an important task of object initialization. As an object is created, Compiler makes a call to the Constructor. The constructor initializes the created object by initializing its members after the storage is provided to that particular object. There is at least one constructor for a class. If there is no explicitly defined constructor in a class, a by default constructor with no parameter comes into existence.

class Test

{

Char c;

public:

Test(); // A Constructor

};

class Test

{

int a;

public:

Test(); //declaration for constructor
Test::Test() // defining Constructor

{

a=1;
}

};

Destructors

Destructors are the special class functions having a special responsibility of destroying the object as that object goes out of its scope. Compiler automatically calls the constructor as an object goes out of the scope.

Destructor has the same syntax as that of constructor with a single difference that tilde sign symbol as ~ is used or destructor declaration. Also a destructor has no argument.

class Test

{

public:

~Test(); // Destructor for class Test

};

Example for demonstration of calling of constructors and destructors

class Test {

```cpp
Test()

}

cout << "A Constructor has been called"; }

~Test()

{

cout << "A Destructor has been called"; }

};

int main()

{

Test test1; // Constructor is called

int a=18;

if(a)

{

Test test2; // again Constructor is Called }// obj2 is out of scope and Destructor is Called for that obj2

} // Destructor is called for the object obj1
```

The Life Cycle of a Class

When creating classes, you want them to be as easy as possible to use. There are a couple of basic operations that any class will likely need to provide support for:

- It needs to initialize itself

- It needs to clean up memory and/or other resources

- It needs to copy itself

All of these are very important for the creation of good data types. We'll use the string to demonstrate this. Strings must be able to initialize themselves, even if they're empty. They should need to rely on external code for that. Once the string is declared, it's immediately available for use. When you're finished with the string, it must be able to clean up because all strings allocate memory. When the string class is used, there is no need for another method to be called for that; the string does it automatically.

Lastly, it must be able to copy from one variable to another, the same way that integers can be copied between variables. Put together, all of this functionality should be made a part of all classes so that they can easily be used.

We'll take these three features, one at a time, and look at how easy C++ makes all of this.

Object Construction

You might have spotted that we didn't have any code for initializing the board in our ChessBoard interface, which was the public section of the code. We can fix that now.

When a class variable is declared, the variable needs to be able to be initialized:

ChessBoard board;

When an object is declared, the code that runs it is called the constructor, and this should set the object up so that it needs no more initialization. Constructors can take arguments too—you saw this with the vector declaration of a particular size: vector<int> v(10);

The vector constructor is called with a value of 10; the new vector is initialized by the constructor so that it can hold 10 integers immediately.

Creating a constructor is nothing more than declaring a function with the same name as the class, with no arguments and no return value. You don't even give it a void value; no type is provided for the return value: enum ChessPiece { EMPTY_SQUARE, WHITE_PAWN /* and others */ };

enum PlayerColor { PC_WHITE, PC_BLACK };

class ChessBoard

{

public:

ChessBoard (); // <-- no return value at all!

PlayerColor getMove ();

ChessPiece getPiece (int x, int y);

void makeMove (int from_x, int from_y, int to_x, int to_y);

private:

```cpp
    ChessPiece _board[ 8 ][ 8 ];

    PlayerColor whosemove;
};

ChessBoard::ChessBoard () // <-- no return value
{
    whosemove = PC_WHITE;

    // empty the entire board to start with and then add in the pieces

    for ( int i = 0; i < 8; i++ )
    {
        for (int j = 0; j < 8; j++ )
        {
            board[ i ][ j ] = EMPTYSQUARE;
        }
    }

    // other code needed for board initialization

}
```

I won't keep showing you the method declarations, but I will continue to show you the class declaration, so that you can see how it all fits together.

Now, note that the constructor above is in the public part of the class. If we didn't make the ChessBoard constructor public, we couldn't create any object instances. Why not? Because, whenever an object is created, the constructor has to be called; if we make it private, it can't be called from outside the class. Objects can only be initialized by calling the constructor and, if you make it private, you won't be able to declare the object.

The constructor gets called on the same line the object is created on:

```cpp
ChessBoard board; // calls ChessBoard constructor
```

Or when memory is allocated:

```cpp
ChessBoard *board = new board; // this calls the ChessBoard constructor as part of the
```

memory allocation

When multiple objects are declared:

ChessBoard a;
ChessBoard b;

Constructors are always run in the order of object declaration—a and then b. As with a normal function, constructors can take one or more arguments, and several constructors can be overloaded by the type of argument if there is a different way of initializing the object. For example, a second constructor could be created for ChessBoard to take the board size: Class ChessBoard

{

ChessBoard ();

ChessBoard (int board_size);

};

The function definition is the same as any class method:

ChessBoard::ChessBoard (int size)

{

// ... code
}

The argument is passed to the constructor like this:

ChessBoard board(8); // 8 is an argument to the ChessBoard constructor

When you use the new keyword, passing an argument looks like you're calling the constructor directly:

ChessBoard *p_board = new ChessBoard(8);

A note on the syntax: although parentheses are used for passing arguments, you can't use them to declare objects that have no-argument constructors.

You can't do this:

ChessBoard board();

The correct form is:

ChessBoard board;

However, you can use parentheses when the new keyword is used:

ChessBoard *board = new board();

This is one of the unfortunate little foibles of C++ parsing because the details are kept very obscure. Just don't use parentheses when you declare objects with a no-argument constructor.
So, what happens if a constructor isn't declared?

If a constructor isn't declared, C++ creates one. It won't take arguments or initialize ints, chars, or any other primitive type, but it will initialize every class field by calling the default constructor for each one.

Generally, you should create constructors just to make sure everything gets properly initialized. As soon as the call constructor is declared, C++ won't generate the default for you. The compiler assumes that you know exactly what you're doing and that you're going to create the constructor you need. In particular, if you create constructors that take arguments, the code no longer has the default constructor unless you declare one specifically.

This can have unexpected consequences. If the automatically generated constructor is used and then you add a non-default one that takes arguments, the code that depends on the automatic constructor won't compile. It's up to you to provide a default constructor because the compiler won't do it for you.

Initialization List and Const Fields
If the class fields are declared as consts, the file has to be initialized in the list:

class ConstHolder

{

public:

ConstHolder (int val);

private:

const int _val;

};
ConstHolder::ConstHolder ()

: _val(val)

{}

Const fields can't be initialized by assigning to them because the fields are set and can't be changed. The only place where the class has not been formed fully is in the initialization list, so immutable objects can safely be set. For the same reason, if a field is a reference, it has to be initialized in the list.

Chapter 8

Library

The Standard Template Library

It's good to be able to write your own data structures. However, as you might have picked up, it isn't all that common. But you didn't work through all the previous content for no reason. You've now learned quite a bit about building data structures as necessary, along with some of the more common structures that you might need when writing a data structure.

However, C++ has a cool feature, a huge library of code that can be reused. It's called the Standard Template Library (STL) and contains the most common data structures, including those built on the binary trees and linked lists. These structures let you specify what data type they will store at the time you create them so that they can be used for anything—structured data, strings, ints, and so on.

Because you have this flexibility, the STL will, in many cases, eliminate the need for you to build structures of your own. Using it can raise your code level in these ways:

You can start thinking of your code in terms of the structures required rather than having to think about building and implementing them yourself.

You have free and easy access to the best data structure implementations, with space use and performance optimized for most problems.

You no longer have to concern yourself with the allocation and deallocation of memory for your data structures.

However, as with everything, using the library comes with its trade-offs:

1. You need to learn the STL interfaces and how they should be used.

2. When you get compiler errors from using the STL, they are incredibly difficult to read and understand.

3. You won't find all data structures in the library.

Thoroughly discussing the STL would require an entire book, so I will only give you an overview of the most common data structures that you will use from the STL.

Vectors

Contained in the STL is an array replacement—the vector. This is much like an array but is resizable automatically; you no longer have to worry about allocating memory and moving the other elements around. However, the vector syntax is different from that of a standard array.

The array syntax is:

int an_array[10];

whereas the vector syntax is:

#include <vector>

using namespace std;

vector<int> a_vector(10);

The vector header file must be included along with the namespace std. This is because, like cout and cin, the vector is a part of the standard library.

Furthermore, when a vector is declared, you must provide the data type being stored in the vector. This is done using angled brackets<>:

Vector<int>

This makes use of a C++ feature known as templates. Vector code is written in such a way that any data type can be stored. All you have to do is tell the compiler what type a vector is storing. What this really means is that we have two types here: the data structure type, which dictates the way data is organized, and the data type held in the data structure. Using a template lets you combine different data structure types with different data types within the data structure.

Finally, when the vector size is provided, it must be placed in a set of parentheses, rather than brackets:

vector<int> a_vector(10);

This is the syntax used for the initialization of certain variable types. For this, the value of 10 is passed to the initialization, called the constructor, and that sets the vector with the size ten. As we go through the guide, you will learn about constructors and objects that have constructors.

Once the vector is created, individual elements can be accessed in the same way as an array:

```
for ( int i = 0; i < 10; i++ )
{
a_vector[ i ] = 0;

an_array[ i ] = 0;

}
```

Calling a Method on a Vector

A vector provides you with much more than the functionality for an array. You can do all sorts of things, such as adding elements beyond the vector end. The vector provides functions to help with this sort of thing, but the function syntax isn't the same as the syntax we used before.

A vector uses a C++ feature known as a method. This is a function that you declare with the data type. Calling methods requires new syntax, as in this example:

a_vector.size();

This is calling a method size on a_vector, and it returns that size. It's much like accessing a structure field but, instead, you're calling a method that goes with the structure. Although something is being done to a_vector by the method, a_vector doesn't need to be passed as an argument to the method. The syntax already knows that a_vector is to be passed as an implicit argument into the size method.

The syntax below

<variable>.<function call>(<args>);

can be thought of as calling a function that goes with the type for the variable. In other words, it's a bit like writing:

<function call>(<variable>, <args>);

In the example:

a_vector.size();

it would be like:

size(a_vector);

Over the next few sections, we will discuss more methods, including ways to declare and use them. For now, it's enough for you to understand that several methods can be called on a vector, using the right syntax. The special method syntax is the only way of making that type of function call. You couldn't, for example, write this: size(a_vector).

Other Vector Features

A vector also makes it very easy to up the number of items held without needing to allocate memory. For example, adding extra items to the vector would be written like this:

a_vector.push_back(10);
This adds another item of 10 at the end of the vector. The vector will take care of the resizing. If you wanted to do this with an array, the memory would have to be allocated, the values copied over, and the new item added in. While a vector does do memory allocation and copying, it does it intelligently, in a manner in which it doesn't have to resize itself every time a new item is added.

However, although you can use push_back to add items to the end of the vector, using the brackets on their own wouldn't be the same. Brackets let you work only with data that has already been allocated, mainly, so that memory allocation isn't done without you being aware.

So, a code like this:

vector<int> a_vector(10);

a_vector[10] = 10; // the last valid element is 9

wouldn't work and would likely result in the program crashing. Not to mention, it's dangerous. Writing it like this:

vector<int> a_vector(10);

a_vector.push_back(10); // add a new element to the vector

works to resize the vector.

Maps
We mentioned maps a little while ago using a value to look up another value. This is common in programming such as when you want to implement an address book for emails where addresses are looked up by name or a program where you look up an account using

an account number or letting user's login to their games.

The STL gives us a map type that lets us specify the key type and value. An example would be a data structure that holds an email address book; this could be implemented as follows:

#include <map>

#include <string>

using namespace std;

map<string, string> name_to_email;

We have to tell the structure that there are two types: string, for the key, and another string for the value, an email address in this case.

One of these maps' helpful features is that when a map is used, the syntax can also be used as an array. If you want to add value, it's much like an array, except the key type is used instead of an integer:

name_to_email["Billy Bunter"] = "billybunter@thisemailaddress.com";

and getting value from a map is much the same:

cout << name_to_email["Billy Bunter"];

You get the simplicity that goes with an array, but with the ability to store whatever type you want. And what's even better is that, unlike the vector, the map size doesn't have to be set before the [] operator is used to add an element.

Removing items from maps is just as easy. Let's say that you fell out with Billy Bunter and want him removed from your address book. You can do that using the erase method:

name_to_email.erase("Billy Bunter");

And you can use the size() method to check what size a map is:

name_to_address.size();

And the empty() method to see if a map is empty:

if (name_to_address.empty())

{

cout << "The address book is empty. I bet you wish you hadn't deleted Billy.";

}

Do not confuse this with how you make a map empty. That's done using the clear() method:

name_to_address.clear();

Because of the consistency in the naming conventions in the STL, the same methods can be used on vectors.

Iterators

As well as data storage and the ability to access elements individually, you can also go through all the items in a data structure. With arrays or vectors, this is just a case of reading each element using the array length. But with maps, because they have both numeric and non-numeric keys, you can't always iterate through using a counter variable.

To get around this, the STL provides us with an iterator, which is a variable that lets you access the elements of a data structure sequentially, even if the data structure doesn't usually have a good way of doing that. We'll start by looking at using iterators with vectors and then move on to using one with the map elements. The idea is that the iterator will store the position in a structure so that you can use that position to access the element. Then you call a method on the iterator to move on to the next element.

Some unusual syntax is required for declaring an iterator. For a vector of integers, it would look like this:

vector<int>::iterator

What this says is that you have the vector and want an iterator that will work for the specific type, which is why we used the ::iterator. Because the iterator marks the position in the structure, the iterator is requested from the data structure: vector vec;

vector<int> vec;

vec.push_back(1);

vec.push_back(2);

vector<int>::iterator itr = vec.begin();

The begin method call will return an iterator that allows you to access the vector's first element.

Iterators can be considered similar to pointers—you speak of element locations in the structure or use it for getting the element. In our case, we read the vector's first element using this syntax:

cout << *itr; // print out the vector's first element

We use the * operator in the same way we do with a pointer; it makes sense when you consider that both store locations.

If you want the next element, the iterator is incremented:

itr++;

This lets the iterator know it needs to move on to the next element.

The prefix operator can also be used:

++itr;

You can see if you have reached the end by making a comparison with the iterator and the end iterator; to do this, call:

vec.end();

If you want code that will loop over a whole vector, it will look like this:

for (vector<int>::iterator itr = vec.begin(); itr != vec.end(); ++itr)
{
cout << *itr << endl;
}

We can loop over a map using a similar approach, but remember that the map doesn't store single values; it stores key/value pairs. When you dereference an iterator, it has two fields—one for key and one for value:

int key = itr->first; // get the key from iterator

int value = itr->second; // get the value from iterator

Below is some code that will display a map's content in an easy-to-read format:

void displayMap (map<string, string> map_to_print)

{

```
for ( map<string, string>::iterator itr = map_to_print.begin(), end =

map_to_print.end();

itr != end;

++itr )

{

cout << itr->first << " --> " << itr->second << endl;

}

}
```

This code is very similar to what is used for iterating over a vector; the only difference is that the map data structure is used along with the first and second fields on the iterator.

Checking a Map for a Value

Sometimes you want to be able to see if a given key has been stored in a map. Let's say you're looking for someone in your address book. You use the find() method to see if the specified value is there and, if it is, retrieve it. An iterator is returned and will either have the object location with the specified key or will be an end iterator if the object wasn't found.

```
map<string, string>::iterator itr = name_to_email.find( "Billy Bunter" );

if ( itr != name_to_email.end() )

{

cout << "How it is to see Billy again. His email is: " << itr->second;

}
```

If you just want to access an element that isn't in the list, use the standard brackets:

```
name_to_email[ "John Doe" ];
```
The map will insert an empty element if the value isn't there already.

Taking Stock

I have only touched on the basics of the STL, but you now have enough information to use some of the foundational types in the library. The vector can be used to replace arrays altogether and, if you don't want to take the time to insert and modify, vectors can also be used in place of linked lists. With the vector type at your fingertips, there are very few reasons why you would want to use an array, and most of those reasons are advanced, such as when you work with the file I/O, which we will discuss at the end of the guide.

The single most useful data type is the map. Using maps makes it easy to write sophisticated programs without having to worry too much about data structure creation. Instead, your attention can be focused on solving problems. In some ways, the map can replace a basic binary tree. Most of the time, you'll want to go into binary tree implementation unless it's for certain performance requirements, or you specifically need a tree structure.
That's where the STL's true power lies: around 80 percent of the time, it gives you the core structures you need, leaving you free to write the code that solves the problems. The rest of the time, you will need the knowledge to build and implement your own data structures.

Some programmers prefer using their code instead of ready-built code. Most of the time, you should NOT use your own data structures; the built-in ones are faster, better, and more complete. However, knowing how to build does give you much better insight into using them.

So, when might you want to use your own structures? Let's say you want a calculator that allows arithmetic expressions to be input by users and then evaluates the inputs with the correct order of operations. An example would be something like 4*9+8/3, evaluated in a way that the division and multiplication are done before the addition.

This kind of structure can easily be thought of in terms of a tree. You could express 4*9+8/3 like this:

+

* /

4 * 8 3

Each node is evaluated in one of two ways:

If the node is a number, the value is returned.

If the node is an operator, the values of the two subtrees are computed, and the operation is performed.

To build a tree like this will require a raw data structure; you can't just use a map. If you only have the STL, you will struggle with this. However, once you understand recursion and binary trees, it becomes a lot easier.

Chapter 9

STL

STL/C++ Standard Library: Containers, Algorithms, Iterators

A common misnomer is to refer to the C++ Standard Library as the STL. The STL, or Standard Template Library, was a standard developed sometime in the past. The C++ Standard Library is a set of templates, many of which were adopted from the STL. The STL is now deprecated, but some extremely useful templates remain.

The Standard Library offers a solution to this problem in the form of containers. Containers manage sets of data of the same type. They include vectors, sets, and lists, among others.

Vectors are much like arrays, except they handle their own storage and size.

Much like with strings, you need to import the vector library in order to use the template:

#include <vector>

Now you're able to access everything regarding vectors.

You declare a template like so:

template<type> variableName;

Let's say we wanted to go back to the earlier example of grades, creating a self-sizing array of grades. We'd declare it like this:

vector<int> grades;

Adding values to a template is simple. You had an element by using the concept of pushing and popping. You push a new value onto the vector, or you pop a value off.

Let's say the first grade you wanted to add was a 55 (poor guy.) We could do this by typing:

grades.push_back(55);

You've now added your first value to the vector.

What if you want to add a set of values? Well, luckily, in C++11, you can also insert a set of vectors in a similar manner to which you initialize a set of values in an array, using the vector.insert() function. This takes two arguments: the index at which you start inserting, and the values you'd like to insert. If you have multiple values in your vector, you can just use vector.end() to find the point of insertion.

grades.insert(grades.end(), { 96, 64, 75, 83 });

In earlier versions, you would need to create a temporary array to insert the data.

int[] tmp = { 96, 64, 75, 83 };

grades.insert(grades.end(), tmp, tmp + 4);

To go through every item in the list, you can use something called a for each loop in other languages. If we wanted to print out every item on the list, we would do the following:

for (int i : grades) {

cout << i << "\n";

}

What if we weren't using a primitive data type but instead an object? There's support for that as well, using something called iterators. By the usage of iterators, you can access methods of the object (such as .length() or .substr() for a string).

If we had a vector of strings, for example, called j and we wanted to print the length of every string therein:

for (string i : j) {

cout << i << " - string length: " << i.length() << "\n"; }

You can do this for any object you create, but we'll talk more explicitly about that when we actually start creating objects in the next chapter.

Another type of container, lists, functions very similarly to vectors. The specific differences between these two are beyond the grasp of a relative beginner. Still, the general rule concerning these two types of containers is that vector is generally the one that should be used unless you have to continually add or erase elements from anywhere other than the end of the container.

There's another type of container called an associative container, which is formatted with a key value and a mapped value. It's generally ordered according to the keys. The most prominent type of associative container is called a map. This is great when you need to associate a set of data with a certain trait. For a real-world example, a store's inventory system could be kept with a map of SKU codes (integers - the key) and item names (strings - the mapped data).

For this, you'd import <map>, and declare a map by:

map<int, string> Inventory;

Inventory[00400030] = "Tamagotchi, blue";

Inventory[00400031] = "Tamagotchi, white";

Inventory[00324359] = "Twilight, DVD";

Inventory[44539294] = "Dark Souls, Xbox 360";
If you wanted to recall a specific element later in your code, you could do something along the lines of:

cout << "Inventory[00400030] is: " << Inventory[00400030] << "\n";

Calling Inventory[00400030] would print out the mapped value, here being "Tamagotchi, blue".

There's another type of associative container called a set. Sets function very similar to maps, except that they don't allow duplicates. Maps allow duplicate values, but not duplicate keys.

Also included in the STL is a library called "algorithms". This library can be used to search, sort, and manipulate element ranges.

Included in these are functions such as equal(), which determines if two sets of elements are equal to each other. There's also transform(), which applies a given function to a given range.

Conclusion

We've come to the end of this book, but this is just the start of a long and fruitful journey for you, a life of programming. With the basics of C++ under your belt, it's time to start your real learning. Start writing those programs! Start implementing data structures and algorithms.

If you enjoyed getting started with C++, there is so much more to learn and do with this wonderful language. Be sure to continue your journey with the second book in the series, which looks at slightly more complex topics while still being beginner-friendly.

C++ programming language is a competitive general-purpose platform which borrows some features of an imperative programming paradigm and can run on any platform. It is referred to as an imperative programming language because it uses step by step processes to achieve its goals. The language uses the concept of object-oriented programming to design and implement complex programs.

The next step is to start writing out some of the different programs and codes that you want with the C++ language.

Even if you are just starting and you just want to spend your time practicing with some of the codes that are in this guidebook, you are still on the right path to seeing some great results with this language.

The more practice you can get along the way with your own coding, the easier it will become to write and create your own codes along the way.

The C++ language is a unique option to work with, one that will help us to create some really powerful and strong codes, and the functionality that is found with this language is unlike any other that we are able to work with.

All of these can be important in some of the codes that you want to write along the way, and when we are done with this guidebook, you should feel ready to handle these and add them to your own codes as well.

As we can see, there is so much that we are able to do when it comes to working with the C++ language.

It is one of the best coding languages out there, whether you are a beginner, or you are someone who has worked with programming and coding for a long time.

When you are ready to start your own coding journey, and you want to be able to create some of your own programs in no time.

C++ programming language utilizes the features of OOP paradigms to develop and execute a program. Some of these features include the use of objects, classes, data abstractions and encapsulation, inheritance, and polymorphism.

The program focuses on the use of data rather than procedures to solve real-world problems. The program tries to eliminate the shortcomings of other conventional programming languages by improving the data security of programs, promoting code reusability, and its ability to use inheritance feature to make an effective program.

Just like many other OOP languages, C++ supports various building blocks. In this tutorial, you have learned how to use various built-in data types when handling simple to complex programs. The data types are divided into primitive data types, derived data types, and abstract data types. You not only learned how to use the primitive or standard data types but also how to declare your own user-defined datatypes.

Enjoy your journey. If you don't have fun while you program, it isn't for you. Make it fun, make it exciting, and make it the reason you want to get up in the mornings.

Good luck with your journey.

C# PROGRAMMING

A complete guide to master C# on your own. Build coding knowledge creating real projects and applications. Transform your passion in a possible job career as a computer programmer.

Michail Kölling

CODING HOOD

Introduction

Welcome to C# programming and thank you so much for picking up this book!
Whether you are a seasoned programmer or a complete novice, this book is written to help you learn C# programming fast.
By the end of the book, you should have no problem writing your own C# programs. In fact, we will be coding a simple payroll software together as part of the project at the end of the book. Ready to start?
Most people are scared of learning a new coding language. They know that it would open a lot of doors for what they would be able to do with their computers, but they worry that coding itself is too hard for them to learn how to work with. If you do not take the proper time to learn a new programming language, the whole process of programming could be difficult. But when it comes to finding a good language that will help you write almost any code that you would like, then it is time to take a look at the C# programming language. This book will take you through some of the basics that come with using the C# coding language so that you can start using it yourself.
What is special about C#?
The first thing that you might want to look into when it comes to a coding language is to understand why C# is so special and why you would even want to learn how to use this particular programming language.
There are many different coding languages out there to choose from, and they all work differently, but you will find that there are a ton of benefits that come with using the C# program, and we will explore some of them inside this guidebook. Even though you are just a beginner, this is a great coding language to work with and will allow you to design so many programs of your own.

While there are many different options available if you want to get started with coding, none are as great to work with like C#. Some of the benefits of going with C# rather than some of the other programming languages include:
- It can utilize a big library

As a beginner, there are a lot of parts of the code that won't be easy for you to learn. You will learn them as you go, but the library that you can use with C# is a great resource that will be of great help to you. You can place these functions into the code without a lot of

hassle being involved. You can even use them to make some changes to the code, so it works the way that you want.
- Automatically disposes of the functions

When you are working with some of the other programming languages, you will have to go through and remove the items that you own. This will take up your time and can be a hassle if you end up missing some of them. Using C# will do all of this work for you to make things faster and easier.
- Easy to learn

C# is widely considered as one of the easiest programming languages you can learn how to work with. While there are a few parts that are more complicated than some other coding languages, this is not a difficult one, and you'll start recognizing different parts of the code pretty quickly as you continue to use it.
- Compatible with Windows computers as well as others

This programming language was originally created to work on Windows computers and help you design a program for them. But it also works well with some other operating systems such as Mac, Linux, and more as long as you download .NET on it. Windows has some great products that are easy to use, especially for beginners, so you will surely get great results once you get started.
- Works with .NET which helps make it easy

This is a program that already comes with the Windows computers, but you can add it to some of the other systems to make C# accessible on these other computers as well.
- Similar to C and C++

This makes it really easy to work with this program and learn the basics before going on to these other programs. Even if you choose to stick with this programming language, you will find that it is powerful enough to do most of the coding that you want and without all the hassle you might experience with some of the other programs.

Ready to dip your toes into the world of C# programming? Let's get started.

Chapter 1

Anatomy of C#

An Introduction to C#

Learning how to code is going to provide you with a ton of benefits along the way. It is going to make it easier to create some of the programs and applications that you would like. It is going to be able to help you further your career with a lot of options that are simple to work with and can allow us to get the best out of these new skills and more money as well. And then there are times when learning how to work with C# is going to be vital because it will allow us to learn more about our computers and how they work.

There are a lot of great options when it is time to work with programming. Many languages have been developed to work on different projects, and sometimes they will work on different types of processes and applications as well. While many of these can be useful based on what you would like to accomplish with the language in the first place, we are going to take a look at how to work with the C# language and how great this one can be for our needs as well. Let's dive into the C# language and how to make this work for our programming needs as well.

The first thing that we need to take a look at here is the C# language. C# is going to be an elegant and object-oriented language that is going to be helpful because it allows programmers to go through and built up a variety of robust and secure applications, all of which are going to run with the .NET Framework. It is possible to work with C# on a variety of projects, including with Windows client applications, Web services, client-server applications, XML database applications, and so much more.

When we are working with the C#, we will find that visually it is going to provide us with an advanced code editor along with some user interface designers that are more convenient to use, a debugger that is integrated, and some other tools that will ensure that it is easier to develop some of our applications based on this language.

The neat thing about the C# language is that it is not only going to be highly expressive and able to help us out with a lot of the more complicated codes that we want to handle, but it is also going to be easy to learn and simple. The curly braces that are part of the syntax of C# are something that is going to look familiar, primarily if you have worked with Java, C++, and C in the past. Developers who are good with these languages will find that it does not take that long to learn C# because of the similarities.

C# is useful because it is going to be able to simplify some of the issues that happen with the C++ language, and it is going to provide us with some powerful features, including value types that can be null, delegates, enumerations, lambda expressions, and direct memory access. None of these are going to be featured available with languages like Java.

In addition to this, you will find that the C# language is going to be able to support some of the generic methods and types, which is going to help us provide increased type safety and performance. It comes with some iterators, which enable the implementers of collection classes to help define custom iteration behaviors that will be easy to use by the client code.

Because this language is going to fit into the category of object-oriented, it is going to be able to support a lot of topics and options that fit into this category, and that we will be able to explore in more detail later on, including polymorphism inheritances, and encapsulation. All methods and variables that we use are going to be encapsulated within these definitions of classes. And a class can inherit directly from one parent class, but it can be set up to implement any number of interfaces.

Methods in this language will be able to override accidental redefinition. When we work with C#, a struct is going to be similar to a lightweight class. This means that it is going to be a stack-allocated type that will be able to come on board and implement interfaces, but it is not going to be able to help us when supporting inheritances.

In addition to some of the basic principles that we just took a look at, you will find that C# is going to make it easier to develop some of the components of software through a few constructs, and these are going to include the following:

It can go through and encapsulate some of our method signatures through the use of delegates. These are going to enable some of the type-safe event notifications that we need.

Properties that are going to be there to work as the accessors for some private member variables.

Attributes, which are going to help provide declarative metadata about some of the different types when the program runs.

LINQ, which is going to stand for a Language Integrated Query. This is useful because it is going to provide us with some built-in query capabilities no matter which sources of data that we are going to work with.

There are a lot of benefits that we can see when it is time to work with the C# language and make it work for some of our needs. It is an object-oriented language that makes it easier to organize and keep things together, and when we can utilize the right IDE, and the .NET platform that we will talk about in a moment, we will be able to utilize it to make the robust programs that we want, without all the complexities of the missing features that happen with other similar languages.

The C# IDE

Before we can get into details about how we can code, and what all we can do with the C# language, we need to get a better idea of what an IDE is about, and why this is such an important part of any coding that we want to do. To start, an IDE is going to be an integrated Development Environment, which is an application we can use to help with application development. In general, this is going to be a GUI or graphical user interface that is designed to make it easier for any developer to build software applications. Many times, these are going to have more of the tools that you need.

Most of the common features, such as data structure browsing, version control, and even debugging, are going to be there to help the developer execute actions without needing to switch back and forth between the different applications that they are working with. This makes it easier to maximize the productivity because we can provide similar user interfaces for the components that are related, and it is going to reduce the amount of time that it takes to learn a new language. Also, depending on the kind of IDE that you choose to go with, it can sometimes just support one language, and other times it will support many languages.

The concept that is behind the IDE is something that evolved from some simple command-based software, which at the time was not as useful as the menu-driven software. Some of the modern IDE's are going to be used more in the context of our visual programming, while the applications in these are going to be created quickly and efficiently by moving the code nodes or the building blocks of programming that will generate the flowchart and structured diagrams. These can then be either compiled or interpreted based on what you hope to get out of the process.

It is important that we take the time to pick out a good IDE, and there are several factors that we need to consider as we work through this. For example, we need to spend time looking at the language support, operating system needs, and some of the costs. Sometimes the IDE that we want to work with will be free, but sometimes these are going to have limited features. It often depends on what you would like to design with that language, and what features are important to your coding needs.

There are several good IDE's out there that are available with the C# language. It is important to do some research and learn more about how these work, and which ones are going to have the features that you would like to get the best results. When you can do this, it is infinitely easier to get the codes written that you would like.

How to Work with C# On Linux and Mac Computers

While C# is considered a Windows language, we can go through and work with it on a Linux and Mac computer as well. We may need to go through a few extra steps, and sometimes a few of the functionalities that we are used to seeing with Windows will be gone, but there are still a lot of benefits that we will see with using this option on other operating systems.

For the Linux system, the MonoDevelop IDE, which is part of the Mono Project, should be just what you need to develop some C# coding on a Linux computer. The easiest way for you to install this is to work with the MonoDevelop package that was developed to work with Ubuntu. Sometimes it is recommended to work with the WinForms toolkit, but this one is not going to be as efficient and easy to work with.

We can also set this language up to work on a Mac computer. The first step is to download an IDE, and Visual Studio Code is usually the best one here because it does work well with Mac and still provides you with the functionality that you need while the simplicity of use is still there. Just visit the Visual Studio website, and then make sure to click on the button to download to Mac. This should download what you need in a zip file.

Once you have been able to get this setup, you can unzip the file and then drag it over to the Applications folder. You should be able to download it from there. The next step is to go through and download the C# extension. You can go into Visual Studio Code and open up the Extensions that you need to make this work.

Now, as you are going through this process, you are going to notice that there is a search bar at the top of the extensions view. This is where you are going to type in "C#". The one that you need is going to be by Microsoft, so make sure to click on this one. Click to install, and you are ready to go.

Remember though this process that the C# language is one that has been developed to work the best with a Windows system. So, if you are on a Windows system, you are more likely to find this language easy to use, and all of the features and functionalities that you are looking for in coding are more likely to be where you put them.

This does not mean that you are not able to get this to work on your Mac computer at all. It merely means that you need to take a few extra steps to get this setup, and it may not be as easy to accomplish some of the things that you would like. Taking some time to get familiar with the C# language on a Linux or Mac computer will be important to helping you get the best results in the process.

How to Get a C# Compiler

There are several compiler options for C# to make the learning easy and smooth for the beginners; in this book, we will use the C# 4.0 version that comes with Visual Studio 2010. Still, it is also possible to use a later or older version. Most of the examples should be able to be compiled and run without any problem.

In case we cannot get the professional version of Visual Studio 2019, a better option is the Community version formerly known as the Express version. This version is free and can be downloaded directly from the Internet; all you need to do is go through a small survey.

To download the C # Visual Studio compiler (Community Version) completely free, you can access the Microsoft portal *here*.

Once we have downloaded the compiler, we can proceed to carry out the installation; this task is very similar to installing any other Windows program, but do not forget to register it within 30 days.

The Development Environment

Once the installation is finished, we can start the program by selecting it from the Windows Start menu. You will see a window that shows the user interface of the application.

The interface is divided into two parts: on the left, we have the search bar named **Open recent**, where we can find links to our most recent projects or solutions by entering keywords. And on the right side of the home page, we will see the options of **Create a new project, open a local folder, Open a project or solution**, and access to code repositories titled *Clone or check out code*. We can also receive announcements and news about software development using C # in this same part of the interface, but as we move through different pages, we will have the editing zones. These editing areas will allow us to edit the program code, edit the user interface, icons, and other resources.

Editing the program code is a very simple activity that should not make us worry if we know how to use any type of text editors such as Microsoft Word or Windows Notepad, and then we can edit the program code. The other editors are also quite user-friendly, but we will not need them to perform the examples shown in this book.

On the right side, we have a window that is known as Solution Explorer, in this same area other windows will appear that give us information about the project we are developing. We will see the details shown by some of these windows a little later in this book.

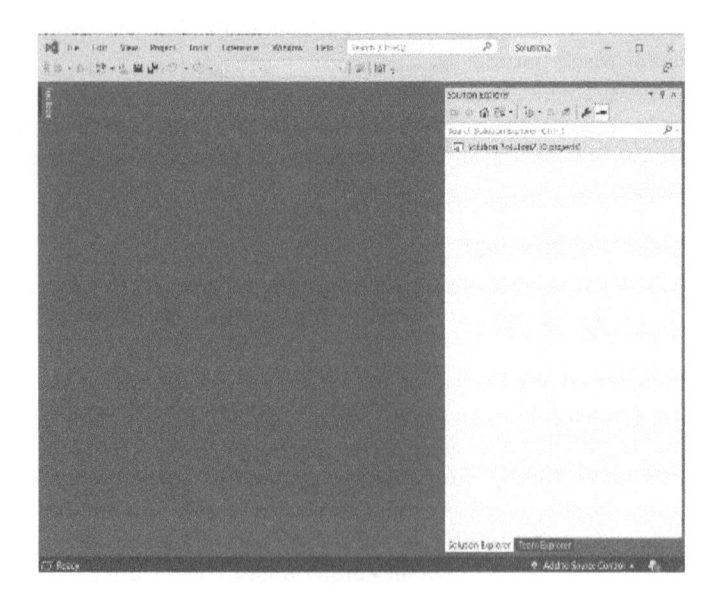

At the bottom, we find another window; in this area, there are usually the sections that the compiler will use to communicate with us. For example, we will see the window that indicates the errors that our program has; at the top of the window, we will see the menus and the toolbars.

If, for any reason, we close the source code editor or it does not appear in our user interface, the easiest way to find it is through the Solution Explorer. Simply go to the document that represents your source code and double click on it, the window with the source code editor will appear.

How to Create Our First Application

To become more familiar with C# Visual Studio Express, it is best to create the first project and work on it. We will generate a small program that sends us a message in the console, and then we will add other elements so that we can explore the different windows of the user interface.

There are mainly two ways to create a new project in Visual Studio 2019. The first one is merely clicking on Create a New Project from the bottom right part of the main visual studio interface.

Then type console in the search box and choose the Console App (.NET Framework) option and click the Next button as shown below.

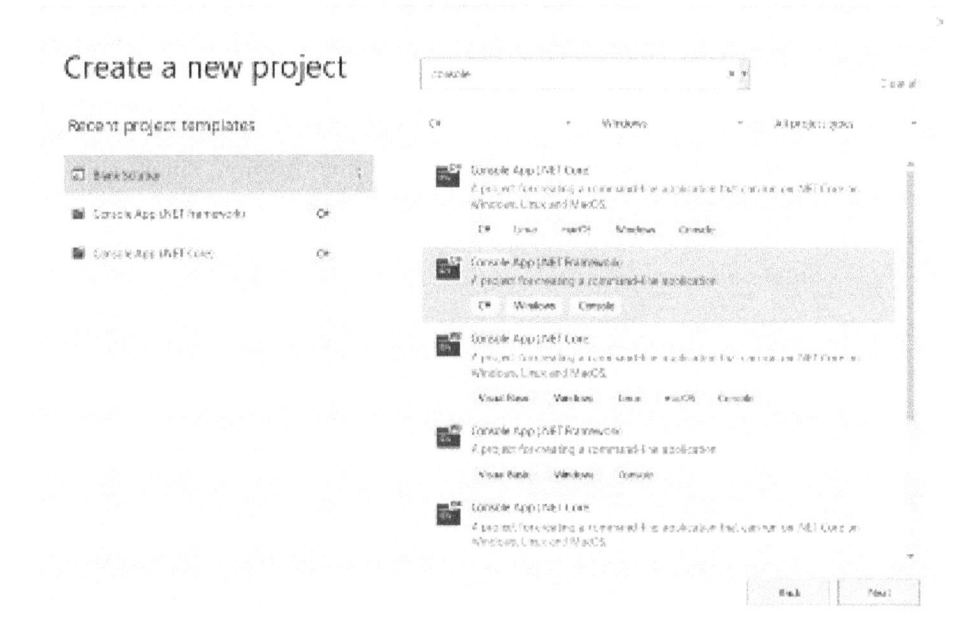

Then configure your new project with its Name and Location, save and hit the create button.
All set, the Visual Studio will instantly create a new project for you.
Alternatively, to create a project, we can select the **File** menu, and after that, we have to click on the option **New Project** as shown in the image below, this way we will see a dialog box that shows some related options.

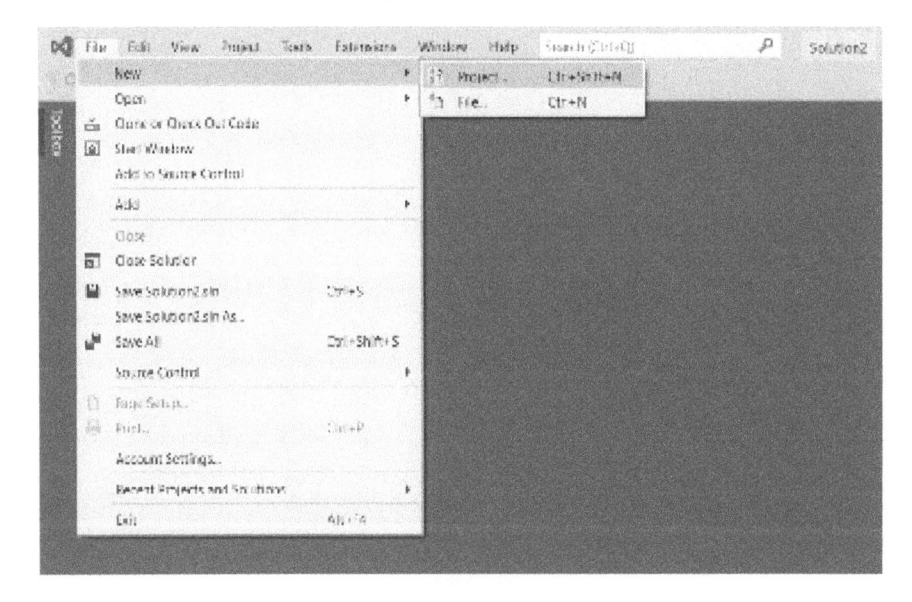

In the right part of the dialog box, the different types of projects that we can create are listed; for this example, we must select the one indicated by the console application and click the next button.

Note: If, for any reason, we close the source code editor or it does not appear in our user interface, the easiest way to find it is through the Solution Explorer. Simply go to the document that represents your source code and double click on it, the window with the source code editor will appear.

On the left side, we will write the name of our project, which in this case, will be MyProject. Each new project we will create will have its name; then, we will simply press the *Create* button.

Configure your new project

Blank Solution   C##

Project name

MyProject

Location

C:\Users\[user]\source\repos

Solution

Create new solution

Solution name ⓘ

MyProject

                                        Back   Create

In a few seconds, C # Express creates the project for us. Visual Studio and the Community version make use of solutions and projects, a solution can have several projects, and for example, Office has different products such as Word, Excel, and PowerPoint. The office is a solution, and each product is a project. The project can be a stand-alone program or a library and can have one or more documents, and these documents can be the source code and additional resources.

We can see that our user interface has changed a bit, the windows already show us information, and we can also see that the skeleton for our program is shown in the editing area.

The Solution Explorer shows us the information about the solution in a logical way if we observe, it is like a small tree. At the root, we find the solution; each project we have in that solution will be a branch, each project, in turn, will also have its divisions. In our case, we see three elements; two of those elements are folders; in one, we keep the properties of the project, and in the other, the references (during this book, we will not use these folders). The fourth element is a document called Program.cs, this represents the document where we save the source code of our application. We see that the extension of the C# programs is .CS.

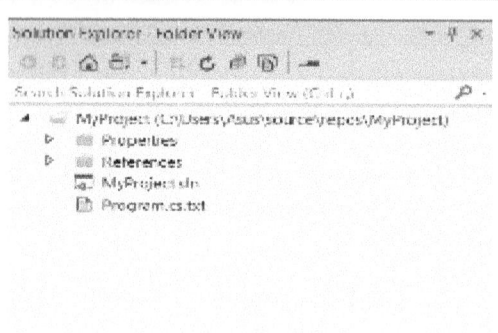

In the editing area, we can see that we have a skeleton so that, from there, we can create our program. To understand what we have, there is necessary to know a concept: namespace. The namespace is a logical grouping; for example, all the code that we can have related to mathematics can be grouped inside the namespace of Math. Another use that namespace has is to solve conflicts with names, for example, let's suppose that we have a massive project and several programmers working on it. Two or more programmers could have created a method that had the same name; this generates a conflict since the program could not know which version to use. The way to solve this is that each programmer has its namespace and refers to the corresponding namespace according to the version we want to use.

The .NET Framework provides us with various namespaces where we have thousands of classes and methods already created for our use. When we want to use the resources found in a namespace programmed by other programmers or by us, we must make use of a C # command known as using.

As we can see at the top of the code, we have several using references to the namespaces that our application needs; if we needed to add more namespaces, we would do it in this section.
Below is defining the proper namespace of our project; this is done as follows:
namespace MyProject
{
}
The namespace we are creating is called MyProject. As we can see, the namespace uses {} as delimiters; this is known as a code block, and everything that is placed between {} will belong to the namespace. This is where it will be necessary to write the code corresponding to our application.

Inside the code block, we find the declaration of a class, C# is an object-oriented language, and that is why it needs us to declare a class for our application. The class has its block of code, and in our application, it will be called Program. The concept of the class will be covered in Chapter 10 of this book.

All programs need a starting point, a place that indicates where the program execution starts, in C#, as in other languages, the starting point is the Main() function; this function also has its code block. Within this function, we will generally place the main code of our application, although it is possible to have more functions or methods and classes. The parts and characteristics of the functions are seen in Chapter 5 of this book.

Now it is time to create our first application. We are going to modify the function code, as shown below. When we are adding the statement inside Main(), we must notice that it happens immediately after the point is placed.

To Compile the Application
Once we finish writing our program, we can carry out the compilation, as we learned earlier; this will generate the assembly that will then be used by the runtime when executed.
To compile the application, we must select the Build menu and then Build Solution or simply pressing by the F6 button. The compiler will start working, and in the status bar, we will see that our solution has been compiled successfully.

To Run the Application
Once the compilation has been successful, we can run our program and see how it works. For this, we have two options: run with debugging and execute without debugging. In the Express version, only the execution with debugging appears, but we can use the execution without debugging with the keys CTRL+F5 or adding it using the tools menu.

The debugger is a program that helps us correct errors at runtime and also logic errors. Preferably we should use the execution without the debugger and make use of the execution with debugging only when we need it. Now we will run our application; for this, we press the CTRL + F5 keys.

When the program is executed, a window appears, we call it console, and it shows the execution of the program. In this way, we can read the message that we had placed in our example.

Since this is a starter book for programming with C#, so we will run all programs on the console. Once you understand the main concepts, you can learn how to program shapes, forms, and graphical interfaces in C#. This type of programming is not difficult, but it requires basic knowledge of object-oriented programming.

How to Detect Errors in a Program

Sometimes we may miswrite the program; when this happens, the application cannot be compiled or executed; if this happens, we must change what is wrong.

Let's write an error intentionally so that we can see how the compiler behaves.
In the program, we have changed '*Console*' to '*Consoli*' and reordered '***namespace***' '***HelloWorld***,' this will cause an error; now, we can try to compile again and see what happens.

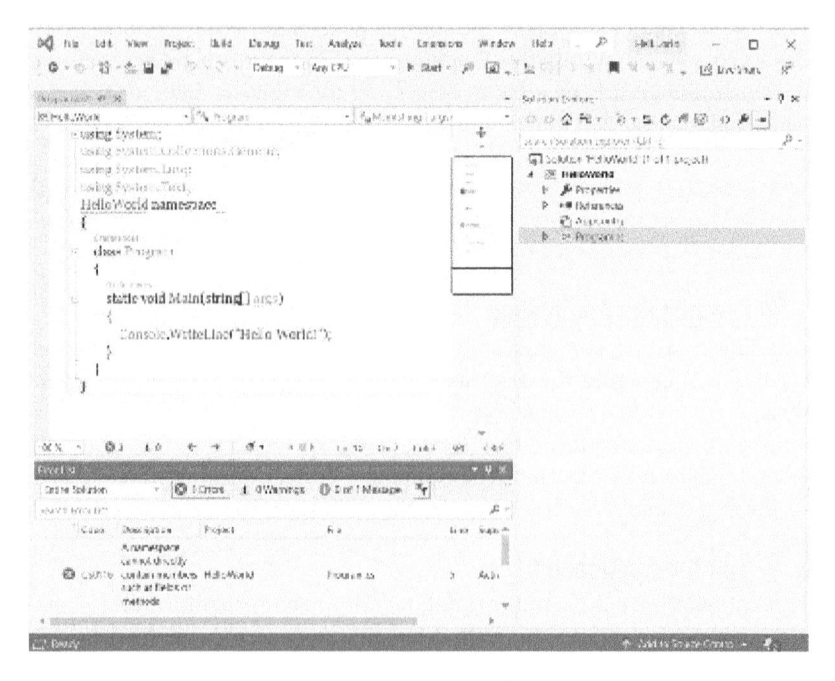

As we already know, the window that appears at the bottom of our user interface is used by the compiler to communicate; we see that a window appears that gives us a list of errors. In the next book, we will learn how to use it; for now, we simply need to know that it is always necessary to solve the first problem in the list and that we can go directly to the error by double-clicking on it.

The Class View

Now we are going to check the class view. In this view, we can get information about the solution we are creating, but unlike Solution Explorer, the information is arranged logically.

With this view, we can quickly find the namespaces of our solution and within them the classes they contain; if we wish, we can see the methods that are in each class. This view not only allows us to observe the logical information but also gives us the possibility to quickly navigate in our code.

In the Community version, there is no previously configured option, so it is necessary to add the command to the View menu.
To show the class view, we must go to the View menu and then select Class View or merely pressing **Ctrl + W, C** a window will appear in our user interface.

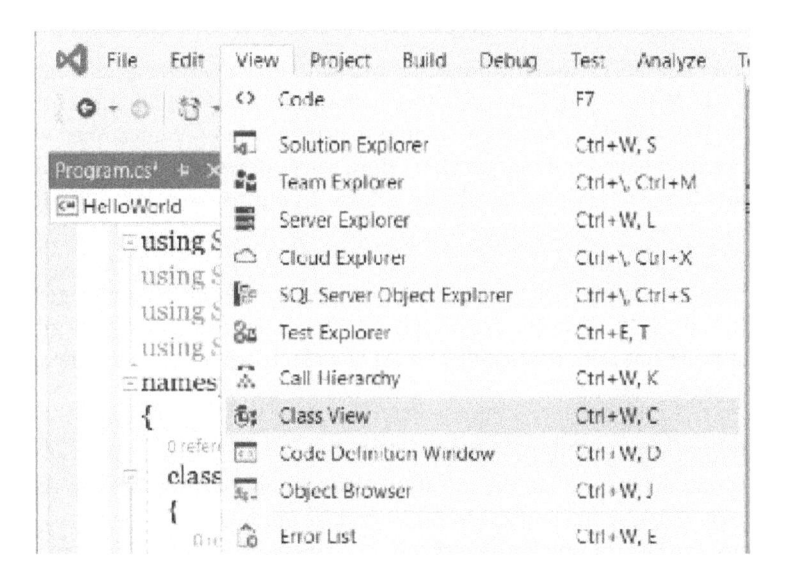

The class view window is divided into two sections; the upper section shows the logical and hierarchical relationship of the elements while the lower section shows the methods that make up a particular class.

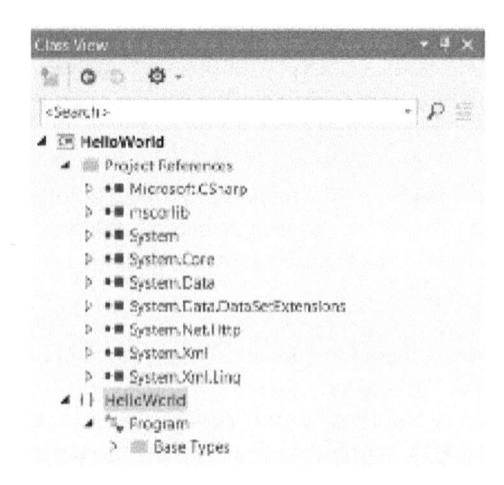

At the root of the tree, we find the project; it will contain the necessary references and the namespaces of HelloWorld. If we had more namespaces, they would appear there, when opening the HelloWorld namespace, we find the classes that are unclaimed within it. In this case, we only have the Program class; if we select it, we will see that at the bottom, the elements that are declared inside are shown. In our example, we have the Main () method.

If, at this time, we double click on any of the methods, we will automatically go to the code where it is defined. Because our program is very small, we may not see the advantage of this, but in programs with thousands of lines of code, being able to navigate quickly is a great advantage.

Configuring Compiler Menus
To add the options we need in the menus, we must follow a series of very simple steps. First, go to the *Tools* menu and select the *Customize* option.
A dialog box appears, in which we must select Commands; in this section, we can choose the menu to which we want to add a command, for example, the View menu.

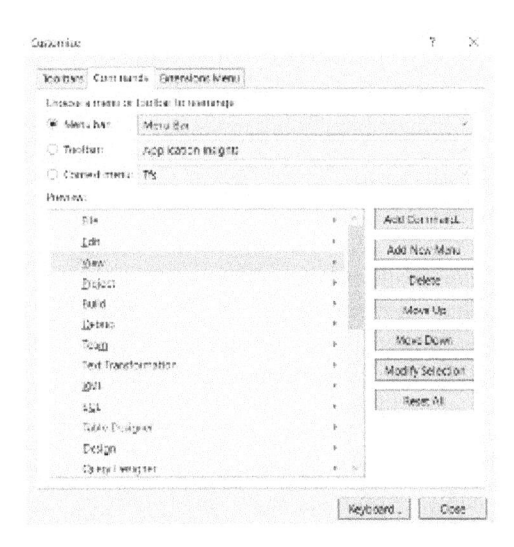

The current menu commands are listed, we click on the area where we want the new command to be inserted and press the Add Command button. With this, a new dialog box appears that shows all the possible commands that we can add classified.

In the categories we select View, and in the Class, View commands, this command is the one that allows us to have the class view. The Start Without Debugging command is in the Debug category.

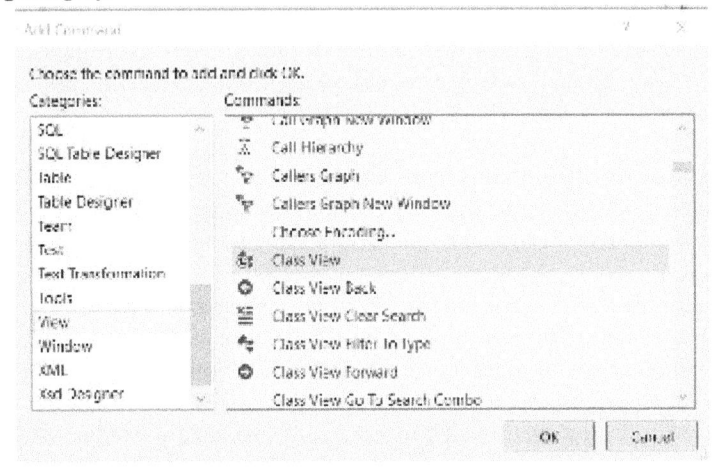

Up to this point, we have analyzed the most important basic concepts needed to understand the operation of .NET. From the next chapter, we will begin with our learning of programming and the C# language.

Main functions of C#

C# is a new object-oriented programming language, which enables programmers to write various applications based on Microsoft .NET platform quickly. Microsoft .NET provides a series of tools and services to maximize the development and utilization of computing and communication fields. It is precise because of the excellent object-oriented design of C# that it is an ideal choice for building various components-whether it is a high-level business object or a system-level application.

The functions of C# are mainly manifested in the following aspects:

- Designing Windows applications.
- Customize the Windows control library.
- Design console application.
- Design smart device applications.
- Design ASP.NET Web application.
- Design ASP.NET Web Service.
- Design ASP.NET Mobile Web Application.
- Customize the Web control library.

ASP.NET is the control and markup developed based on C#. In the field of intermediate language, C# is the most affinity language, which has the main characteristics of C language and Java language, powerful function library, and convenient template, and is one of the ideal languages at present.

Main features of C#

C# language almost combines the advantages of current high-level languages and has the following main features.

Concise grammar

C# language, like Java language, uses unified operators, eliminates complex expressions and pseudo-keywords in C++ language, and makes it described in the simplest and most common form.

Excellent object-oriented design

C# language is designed according to the idea of object-oriented, so it has all the characteristics that object-oriented should have, namely encapsulation, inheritance, and polymorphism.

C# language only allows single inheritance, that is, a class does not have multiple base classes, thus avoiding the confusion of type definition. In C# language, each type is an object, so there are no concepts such as global function, global variable, and global constant. All constants, variables, attributes, methods, indexes, events, etc. must be encapsulated in classes, which makes the code more readable and reduces the possibility of naming conflicts.

Close integration with the Web

In C#, complex Web programming and other network programming are more like operating local objects, thus simplifying large-scale and in-depth distributed development. Components built-in C# language can serve the Web conveniently and can be called by any language running on any operating system through the Internet.

Complete security and error handling

Language security and error handling ability is an important basis to measure whether a language is excellent or not.C# language can eliminate many common mistakes in software development and provide complete safety performance, including type safety. By default, the codes downloaded from the Internet and Intranet are not allowed to access any local files, and resources.C# language does not allow uninitialized variables and provides functions such as convenient check and overflow check. The garbage collection mechanism in memory management greatly reduces the burden of memory management for developers.

Version processing technology

C# language has built-in version control functions, such as handling function overloads and excuses, and feature support, etc., to ensure convenient development and upgrade of complex software.

Flexibility and compatibility

In the managed state, C# language can't use a pointer but uses Delegate to simulate the function of a pointer. If you need to use pointers in classes or methods of classes, you only need to declare these contents unsafe. Also, although the C# language does not support multiple inheritances of classes, it can be realized by inheriting interfaces.

Compatibility means that C# language allows interoperation with API with C/C++ language style that needs to pass pointer type parameters and allows interoperation between C# language components and other language components.

Better correspondence between business process and software implementation

If an enterprise's business plan is to be put into practice, a close correspondence must be established between the abstract business process and the actual software implementation. It is difficult to do this in most languages.

C# language allows type-defined, extended metadata. These metadata can be applied to any object. Project builders can define domain-specific attributes and apply them to any language element, such as classes and interfaces.

In a word, C# is a modern object-oriented language, which enables programmers to create solutions based on Microsoft .NET platform quickly and conveniently. This framework enables C# components to be easily transformed into XML network services so that applications of any platform can call it through the Internet.C# enhances the efficiency of developers. At the same time, it is committed to eliminating errors in programming that may lead to severe results. C# enables C/C++ programmers to develop the network quickly while maintaining the strength and flexibility that developers need.

With this, we have completed a brief introduction to the importance and applications of C# with varied examples. In the next chapter of this book, we will talk about the C# development environment that is essential for developing a well-versed C# desktop window application. Follow along to know more about it.

Chapter 2

Data Type

Data Types

In the C programming language, data types allude to a framework utilized for pronouncing variables or functions of distinctive data types. The type of a variable decides the amount of space it will take and how the bit pattern saved in it is utilized. The fundamental classification used for data types is given below.

- Basic Types: They are number-crunching sorts and comprises of the two following types: (a) integer sorts and (b) floating-point sorts.

- Enumerated sorts: They are again number sorts and are utilized to characterize variables that must be allocated discrete number values all through the system.

- Void type: The keyword void demonstrates that no value can be assigned.

- Derived sorts: They incorporate (a) Array, (b) Pointer, (c) Union, (d) Structure and (e) Function.

The array and structure data types are also referred to as aggregate types. The type of function defines the kind of value the function will return upon termination. We will discuss the essential data types in the accompanying segments.

Integer Types

Here is a list of data types that follow under this category. Also, the storage space occupied by them and their range is also specified for your reference.

- char
  o Allocated Memory: 1 byte o Range: -128 to 127 or 0 to 255

- signed char o Allocated Memory: 1 byte o Range: -128 to 127

- unsigned char o Allocated Memory: 1 byte o Range: 0 to 255

- int

o Allocated Memory: 2 or 4 bytes o Range: -32,768 to 32,767 or -2,147,483,648 to 2,147,483,647

- short

o Allocated Memory: 2 bytes o Range: -32,768 to 32,767
- unsigned short o Allocated Memory: 2 bytes o Range: 0 to 65,535

- unsigned int o Allocated Memory: 2 or 4 bytes o Range: 0 to 65,535 or 0 to 4,294,967,295

- long

o Allocated Memory: 4 bytes o Range: -2,147,483,648 to 2,147,483,647

- unsigned long o Allocated Memory: 4 bytes

Range: 0 to 4,294,967,295

To get the precise size of a variable or data type, you can utilize the sizeof operator. The declarations sizeof(<data type>) yields the size of the data type or variable in bytes. Given below is an example, which illustrates the concept, discussed below: #include <limits.h> #include <stdio.h> int main() {

printf("Data type char (size in bytes): %d \n", sizeof(char)); return 0;
}
Upon compilation and execution of this code, you must get the following output: Data type char (size in bytes): 1

Floating Point Data Types
Here is a list of data types that follow under this category. Also, the storage space occupied by them, their range, and precision value are specified for your reference.
- float
o Allocated Memory: 4 byte o Range: 1.2e-38 to 3.4e+38

o Precision: 6 decimal places • double
o Allocated Memory: 8 byte o Range: 2.3e-308 to 1.7e+308

o Precision: 15 decimal places • long double o Allocated Memory: 10 byte o Range: 3.4e-4932 to 1.1e+4932

o Precision: 19 decimal places The header file named float.h characterizes macros that permit you to utilize these data type values. The following code will allow you to find the exact amount of allocated memory in bytes on your system for the concerned data type.

```
#include <float.h> #include <stdio.h> int main(){

printf("Allocated Memory for float : %d \n", sizeof(float)); printf("Precision: %d\n",
FLT_DIG ); printf("Max Range Value: %E\n", FLT_MAX ); printf("Min Range Value:
%E\n", FLT_MIN ); return 0.

}
```

Upon compilation and execution of this code, you must get the following output: Allocated Memory for float: 4

Precision: 6

Max Range Value: 3.402823E+38

Max Range Value: 1.175494E-38

The void Type
The void data type points out that no value is accessible. It is utilized as a part of three sorts of circumstances:

• Void returned by a function. You must have commonly noticed the use of the data type void as the return type of function. If not, you will see extensive use of the same as you move forward in your experience with C. The void data type signifies that the function will not return anything. Example of such an implementation is: void print(int)

• Function Arguments as void There are different functions in C, which don't acknowledge any parameter. A function with no parameter can acknowledge as a void. Example of such an implementation is: int print(void)

• Pointers to void A pointer of sort void * signifies the location of a variable. For instance, consider the following declaration: void *malloc(size_t size); This function returns a pointer to void. In other words, this function can return a pointer to a location of any type.

You may not be able to comprehend the use and meaning of the void data type in entirety right now. However, as you move forward, you will find it easier to relate to and use this data type in your code..

C#

Chapter 3

Operators Variable

Variables

A variable is simply a name that is assigned to a storage area. Our programs are capable of manipulating such a storage area. Each C# variable is associated with a type that determines the amount of space allocated to that variable in the memory. This also determines the kind of operations that can be applied to the variable. For example, you cannot multiply string variables.

In C#, the following syntax should be used in the variable declaration:

<data_type> <variable_list>;

The data_type in the above syntax must be a valid data type in C# like int, String, float, etc. The variable_list can be many variable names separated using commas.

Below are examples of valid variable declarations:

int a, b, c; char c, d; float x, salary; double y;

A variable can be initialized during its declaration.

For example: int x = 40;

Variable Initialization

Variable initialization refers to the process of assigning a value to a variable. In C#, this is done using an equal sign then followed by constant expression. Here is the general syntax for initialization: variable_name = value; As we had stated earlier, variables can be initialized during their declaration. Here are more examples:

```
int x = 1, y = 6; /* initializing x and y. */
byte f = 22; /* initializes f. */
double pi = 3.14159; /* declaring an approximation of pi. */
char a = 'a'; /* the variable a has a value of 'a'. */
```

Variable initialization should be done correctly; otherwise, the program may give unexpected results.

For example: using System; namespace VariableDeclaration {
class MyProgram {
static void Main(string[] args) {
short x; int y ; double z; /* actual initialization of variables */
x = 5; y = 10; z = x + y; Console.WriteLine("x = {0}, y = {1}, z = {2}", x, y, z);
Console.ReadLine(); }
}
}
The code returns the following:

```
x = 5, y = 10, z = 15
```

Getting User Input
The Console class of the System namespace has a method named **ReadLine()** that allows us to get input from the user. The user's input is read and stored in a variable. The following example demonstrates how to use this method: using System; namespace UserInputApp {
class MyProgram {
static void Main(string[] args) {
string firstName = **"Nicholas"**; string lastName = **"Samuel"**;

Console.WriteLine(**"Your name is: "** + firstName + " " + lastName);

Console.WriteLine("Please enter a different first name:"); firstName = Console.ReadLine();

Console.WriteLine(**"Your new name is: "** + firstName + " " + lastName);

Console.ReadLine(); }
}
}

When prompted to enter a new value for the first name, do so and hit the enter key. You will realize that your name has changed.
Operators
Operators are symbols that instruct the compiler to perform a certain logical or mathematical manipulation. C# has a wide variety of operators. Let us discuss them.

Arithmetic Operators
These are used for performing various mathematical operations. They can be used as shown in the following example.

```csharp
using System; namespace AithmeticOperatorsApp {
class MyProgram {

static void Main(string[] args) {

int x = 31; int y = 10; int z; z = x + y; Console.WriteLine("1: - z equals to {0}", z); z = x - y;
Console.WriteLine("2: - z equals to {0}", z);

z = x * y; Console.WriteLine("3: - z equals to {0}", z); z = x / y; Console.WriteLine("4: - z equals to {0}", z); z = x % y; Console.WriteLine("5: - z equals to {0}", z); z = x++; Console.WriteLine("6: - z equals to {0}", z); z = x--; Console.WriteLine("7: - z equals to {0}", z); Console.ReadLine(); }
}
}
```

Most of the above operators are well known except for a few of them. The ++ is the increment operator, and it increases the value of the variable by 1 for each iteration. The – is the increments operator, and it decrements the value of the variable by 1 after each iteration. The % is known as the modulus operator, and it returns the remainder after division.

The code returns the output shown below:

```
1: - z equals to 41
2: - z equals to 21
3: - z equals to 310
4: - z equals to 3
5: - z equals to 1
6: - z equals to 31
7: - z equals to 32
```

Logical Operators
C# also has a number of logical operators. Let us discuss them briefly:
Logical AND (&&)- the condition is true if both operands are nonzero.

Logical OR (||)- the condition is true if any of the operands is nonzero.

Logical NOT (!)- the operator reverses the logical state of an operand. If the condition is true, this operator will make it false.

The following example demonstrates how to use the above operators: ***using*** System;
namespace LogicalOperatorsApp {
class MyProgram {
static void Main(string[] args) {
bool x = ***true;*** bool y = ***true; if*** (x && y) {

```
Console.WriteLine("1: The condition is True"); }
if (x || y) {
Console.WriteLine("2; Thecondition is True"); }
/* let us change the values of x and y */
x = false; y = true; if (x && y) {
Console.WriteLine("3: The condition is True"); } else {
Console.WriteLine("3: The condition is not True"); }
if (!(x && y)) {
Console.WriteLine("4: The condition is True"); }
Console.ReadLine(); }
}
}
```

The code returns the following output:

```
1: The condition is True
2; Thecondition is True
3: The condition is not True
4: The condition is True
```

Assignment Operators

C# assignment operators can be used as demonstrated in the following example:

```
class MyProgram {
static void Main(string[] args) {
int x = 31; int z; z = x; Console.WriteLine("1: = The value of z is = {0}", z);
z += x; Console.WriteLine("2: += The value of z is = {0}", z);
z -= x; Console.WriteLine("3: -= The value of z is = {0}", z); z *= x; Console.WriteLine("4: *= The value of z is = {0}", z);
z /= x; Console.WriteLine("5: /= The value of z is = {0}", z); z = 100; z %= x; Console.WriteLine("6: %= The value of z is = {0}", z); z <<= 2; Console.WriteLine("7: <<= The value of z is = {0}", z); z >>= 2; Console.WriteLine("8: >>= The value of z is = {0}", z); z &= 2; Console.WriteLine("9: &= The value of z is = {0}", z); z ^= 2; Console.WriteLine("10: ^= The value of z is = {0}", z); z |= 2; Console.WriteLine("11: |= The value of z is = {0}", z);
Console.ReadLine(); }
}
}
```

The code returns the following result:

```
1:  =   The value of z is = 31
2:  +=  The value of z is = 62
3:  -=  The value of z is = 31
4:  *=  The value of z is = 961
5:  /=  The value of z is = 31
6:  %=  The value of z is = 7
7:  <<= The value of z is = 28
8:  >>= The value of z is = 7
9:  &=  The value of z is = 2
10: ^=  The value of z is = 0
11: |=  The value of z is = 2
```

C#

Chapter 4
String and list

List

The List<*T*> is the collection class that is most frequently used in all collection classes. It is a type whose main characteristic is the element accessibility by index. As mentioned before, List class comes under the System. Collections.Generic namespace. The List is a class that provides a lot of methods for the list and element manipulation. Some of them are searching, sorting, etc. It is used for creating a collection of many different types. It could be a collection of integers, strings, and many more. For the reference types, the List allows null value as an element. It is also possible to insert duplicate values inside any List collection. This class can use both the ordering and equality comparer. Arrays are similar to the lists, but the lists can resize dynamically, but arrays cannot.

String

String data type in C# is a type that represents text and works with text variables. It is an array of characters. Declaring a string data type is done by string keyword. When a string keyword is being used, it means that it is referring to the 'System.String' class (classes will be explained later in the book). There are many extension methods which could be used over some string variable. Some of the most popular are Concat - which is concatenating two strings, Contains - which determines whether the string contains the given string from the parameter as passed value, Equals - which determines whether two strings have the same value, etc. The default string data type is an empty string.

Chapter 5

Syntax

Loop structure statement

In real life, we often repeat the same thing many times. For example, when you scrape your eyes in the fourth round of eye exercises, you will repeat the action of scraping your eyes. When playing table tennis, you will repeat the swing and so on. In C#, there is a statement that can execute the same code block repeatedly, which is called a circular statement. Loop statements are divided into three types: while loop statement, do... while loop statement and for loop statement.

While loop statement
The while loop statement is somewhat similar to the conditional judgment statement mentioned in the above section, which determines whether to execute the execution statement in {} according to the conditional judgment. The difference is that the while statement will repeatedly judge the conditions. As long as the conditions are true, the statements in {} will be executed until the conditions are not true and the while loop ends.

The syntax structure of the while loop statement is as follows:
While (cyclic condition)
{
execute statement
...
}
In the above syntax structure, the execution statement in {} is called the loop body, and whether the loop body is executed depends on the loop condition. When the loop condition is true, the loop body will execute. When the loop body finishes executing, it

will continue to judge the loop condition. If the condition is still true, it will continue to execute, and the whole loop process will not end until the loop condition is false.
Next, a case is used to print natural numbers from 1 to 4, as shown in Example below.

Example Program11.cs
```
public class Program11
{
public static void // This is where you need to enter the logic
{
int firstsample = 1;//define the variable 'firstsample' with the initial value of 1
while (firstsample <= 4)

// cycle condition
{
Console.Enteroptions("firstsample = " + firstsample);
//If the condition holds, print the value of Firstsample.
firstsample++;
//firstsample for self-increment
}
Console.ReadKey();
}
}
```
The running results are shown on the computer screen.

For example, the initial value of x is 1, and if the loop condition x <= 4 is met, the loop will be repeatedly executed, and the value of x will be printed, and x will increase automatically. Therefore, the values of x in the printed results are 1, 2, 3, and 4, respectively. It is worth noting that the code in line 9 of the example is used to change the value of variable x in each loop to achieve the purpose of changing the loop condition finally. Without this line of code, the whole loop will go into an infinite loop and never end.

Do-while loop statement

The function of do-while loop statement is similar to that of while loop statement. The difference between them is that while the statement needs to judge the loop condition first and then decide whether to execute the code in {} according to the result of the loop condition. In contrast, the do-while loop statement needs to execute the code in {} once before judging the loop condition.

Its syntax structure is as follows:
```
do
{
execute statement
...
} while (cyclic condition);
```

In the above syntax structure, the execution statement in {} after the keyword do is a loop body. The do-while loop statement puts the loop condition behind the loop body. This means that the loop will execute unconditionally once, and then decide whether to continue execution according to the loop conditions.

Next, use the do-while loop statement to rewrite Example, as shown in the below Example.

Example Program12.cs:
public class Classicprogram
{
public static void // This is where we usually enter the options
{
int secondinstance = 1;
//define the variable 'secondinstance' with the initial value of 1
do
{
Console.Getresults(" firstinstance= " + secondinstance);
//print the value of 'firstinstance'
secondinstance++;
//increase the value of 'firstinstance' by itself
}
while (firstinstance <= 4);
//cycle conditions
Console.Scanprogram();
}
}

The running results will be shown on the computer screen.

Examples above have the same running results, which shows that a do-while loop and while loop can achieve the same function. However, there are differences between these two statements in the process of a program running. If the loop condition does not hold at the beginning of the loop statement, the loop body of the while loop will not be executed once, but the loop body of the do-while loop will still be executed once. If the cycle condition x<=4 in the example is changed to x < 1, the example 2-12 will print x=1, while the example 2-11 will print nothing.

for loop statement

The while loop and the do-while loop was explained in the previous section. In program development, another kind of loop statement is often used, that is, for loop statement, which is usually used when the number of loops is known, and its syntax format is as follows.

For (initialization expression; Circulation conditions; Operation expression)
{
execute statement
...
}

In the above syntax structure, the () after the for keyword includes three parts: initialization expression, loop condition, and operation expression, with "; "between the separation, the execution statement in {} is a loop body.

Next, the initialization expression, loop condition, operation expression, and loop body are represented by 1, respectively, and the execution flow of the for loop is analyzed in detail by serial number.

```
for(1; 2;3)
{
4
}
```
Next, sum natural numbers 1~4 through a case, as shown in Example below.
Example Program13.cs:
```
public class Classicprogram
{
public static void // This is where we usually enter the options
{
int secondinstance = 1;
//define the variable 'secondinstance' with the initial value of 1
do
{
Console.Getresults(" firstinstance= " + secondinstance);
//print the value of 'firstinstance'
secondinstance++;
//increase the value of 'firstinstance' by itself
}
while (firstinstance <= 4);
//cycle conditions
 for (int first = 1; first <= 4;The value of first++)
 //first will change from 1 to 4
{
sum += first;
//realize the accumulation of sum and i.
}
```
In example, the initial value of variable I is 1, and if the judgment condition i<=4 is true, the loop sum+= I will be executed. After the execution is finished, the operation expression i++, i will be executed, and the value of I will become 2. Then continue to judge the condition, and start the next loop until i=5 and the condition i<=4 is false, end the loop, execute the code after the for loop, and print "sum=10".

Chapter 6
Classes

Classes and Objects
In C# programming language, every component is associated with classes and objects, and these represent the basic concepts of object-oriented programming. A class is something like a prototype from which the user creates objects. A class represents a single, unique unit with all of its members, attributes, and functionality.

The objects are real-life, in-memory entities. When instantiated, they allocate some memory space and have reference to it. Every object created must be of a class type.

Class
Classes represent the unique programming components inside any object-oriented software. The **class** is defined with its access modifier, class keyword, and a unique name. Inside some class, there could be multiple members such as fields, properties, and methods. The default access modifier of a class is *internal*. This means that if the access modifier is not specified, then that particular class would be treated as *internal*. An example of a class is written below:

```csharp
public class TShirt
{
    private string color;
    public string Color
    {
        get { return color; }
        set { color = value; }
    }

    public TShirt(string color)
    {
        this.color = color;
    }

    public string About()
    {
        return "This t-shirt is " + Color + " color.";
    }
}
```

This class contains its access modifier – public, which contains the keyword class; it has a unique name, *TShirt*. Inside this class, there are multiple members defined. There is one private field *color*, one public property *Color*, one constructor defined (with one parameter), and one public method *About*.

Apart from the usual classes, there is one special type of class - the *abstract class*. An abstract class is a special type of class in which an object cannot be instantiated. Abstract classes are mostly used to define a base class in the hierarchy, and it is also known as an incomplete class. The abstract classes typically represent a base class. It is designed to have derived classes implementing the abstract definitions. In the abstract classes, there could as well be *abstract methods* or non-abstract methods. The abstract methods, as well as every class member that is marked as abstract, must be implemented in the derived class. The classes that are marked with the abstract keyword have the purpose of providing the prototype for the derived classes. An abstract class can have a constructor implemented. An example is provided below:

```csharp
class Program
{
    static void Main(string[] args)
    {
        Animal animal;
        animal = new Elephant();
        animal.LegsNumber();
        animal = new Pigeon();
        animal.LegsNumber();
    }
}

public abstract class Animal
{
    public abstract void LegsNumber();
}
public class Elephant : Animal
{
    public override void LegsNumber()
    {
        Console.WriteLine("Elephant has four legs.");
    }
}
public class Pigeon : Animal
{
    public override void LegsNumber()
    {
        Console.WriteLine("Pigeon has two legs.");
    }
}
```

In the provided example, we can see one abstract class with the name Animal. In this class, there is only one member, and that is an abstract void method LegsNumber(). When the method is marked as abstract, it means that every class derived from that class, as in this case, an Animal class, must have an override implementation of that method. Below the abstract class, there are two more classes, which are the 'normal' classes. Both classes have the override implementation of a LegsNumber() method from the Animal class. That is because both the Elephant and Pigeon classes inherit from the Animal class. Inheritance will be explained in detail in a later chapter, but for now, the ':' symbol next to the Elephant and Pigeon class means that they inherit from the Animal class. In this case, this means that they must have their implementation of the abstract method from the base class. In the Main method of a Program class, there is a declaration of an Animal type variable named *animal*.

Note that there is no instantiation of the Animal class because that is not possible, and the reason is that the Animal class is an abstract class - so it cannot be instantiated. In the next line, the program is instantiating an object of Elephant class into the Animal type variable. This is allowed because Animal is the base class of the Elephant class. After it, there is a call to the LegsNumber method from the animal variable. Furthermore, since the animal is instantiated to be an Elephant type, this call executes the LegsNumber method from the Elephant class. The next thing is the instantiation of the Pigeon object and assigning it to the same animal variable. It ends up with calling the LegsNumber method again from the animal object. This time implementation of the LegsNumber from the Pigeon class is used. The console output will look like this:

```
Elephant has four legs.
Pigeon has two legs.
```

Object

Objects in C# represent the real entities in the system, with all of their class type characteristics and features. They are located somewhere in memory and have reference pointer to it. Whenever the keyword new is used, an object in memory is created. Objects must be of a class type. Every class which is created in C# programming language is derived from the System.Object class (inheritance will be explained in later chapters). This means that there is a built-in class Object in C #, and every object of any type is derived from that class. Every object created, besides its own functionalities, also has the methods that are available from their parent class Object. These methods are: Equals, ToString, Finalize, and GetHashCode. This could be checked in the example:

```
using System;

namespace ClassesAndObjects
{
    class Program
    {
        static void Main(string[] args)
        {
            var tshirt = new TShirt("blue");
            tshirt.
        }
    }

    public class TShirt
    {
        private str
        public str
        {
            get { return color; }
            set { color = value; }
        }
```

About   string TShirt.About()
Color
Equals
GetHashCode
GetType
ToString

This example is using the class TShirt, which was shown in the previous example in Class subchapter. In the Main method, an object of the TShirt class type is instantiated. This object is referenced by the tshirt variable. In the second line, you can see the available methods and properties that could be executed over the tshirt object. Besides Color property and the About method that are part of the TShirt class, there are also four methods from the parent class Object (Equals, GetHashCode, GetType, ToString).

Interface

Another important component in object-oriented programming is an interface. An interface represents multiple declarations of some functionalities. A class can implement one or more interfaces, but it can only inherit from one class or abstract class. This reveals another C# programming language characteristic - it does not support multiple inheritance. Classes that implement some interface must provide a full definition of all interface members. In the interface usage, there is no manipulation with access modifiers as all the interface members are considered to be public. This is because interface existence is all about its functionality to be implemented by other classes. If some class must implement interface members, it means that the interface members must be public to be implemented by other classes. The interface is an object-oriented component that has declarations, but it cannot have definitions - implementations. If you try to insert some implementations in the interface, the compile-time error will appear. The interface can contain properties and methods - everything that can have the implementation. It cannot have fields and a constructor - because it is not a class, and it cannot be instantiated. On implementation of the interface, the class must implement all of the interface members. As mentioned, multiple inheritance is not supported in C#, but it can be achieved with interface usage since a class can implement multiple interfaces. Example:

```csharp
namespace ClassesAndObjects
{
    interface IPerson
    {
        bool IsRunning { get; set; }
        bool IsStanding { get; set; }
        bool IsSitting { get; set; }
    }
}
```

Here we have one interface defined. It has three boolean properties inside. A class that will implement this interface must have the definition of all these interface members.

```csharp
namespace ClassesAndObjects
{
    class Person : IPerson
    {
        private string _firstName;
        private string _lastName;

        public Person(string firstName, string lastName)
        {
            _firstName = firstName;
            _lastName = lastName;
        }

        public bool IsRunning { get; set; }
        public bool IsStanding { get; set; }
        public bool IsSitting { get; set; }
    }
}
```

The program contains one class, which is a Person class. This class implements the IPerson interface that has been declared. This class has two private string fields, which are _firstName and _lastName. It also has a constructor where those field values are assigned. Besides that, this class has the definition of three members of the IPerson interface – the IsRunning, IsStanding, and IsSitting properties. The compiler is okay with this, as there are no errors. But, if we remove any of these three properties from a Person class, the compiler would report an error. For example:

```
namespace ClassesAndObjects
{
    class Person : IPerson
    {
        private string                interface ClassesAndObjects.IPerson
        private string
                                     'Person' does not implement interface member 'IPerson.IsRunning'
        public Person(s                Show potential fixes (Alt+Enter or Ctrl+.)

            _firstName = firstName;
            _lastName = lastName;
        }

        //public bool IsRunning { get; set; }
        public bool IsStanding { get; set; }
        public bool IsSitting { get; set; }
    }
}
```

Here, we have commented on the IsRunning property that is inside the Person class. The compiler is reporting an error that the Person class does not implement that property, and the program could not build. After this, we will uncomment the property and return to the valid state of a program. Let us create a Program class with the Main method inside. There, we will create an instance of a Person class and run the program to check if everything is working well.

```
namespace ClassesAndObjects
{
    class Program
    {
        static void Main(string[] args)
        {
            Person person = new Person("Peter", "Parker");
        }
    }
}
```

Now, run the program:

Everything went smoothly.

Chapter 7

LINQ, queries, operators – XAML

LINQ

Language integrated query, better known as LINQ, represents query syntax in C# programming language, and it is used for retrieval and filtering of the different sources of objects. LINQ produces a single querying interface for various kinds of objects and variables. This query syntax is integrated into C# programming language, as well as in Visual Basic. For better understanding, we will provide an example to explain it further. Therefore, you have SQL as a Structured Query Language. SQL is used to get and save some data into the database. It is a language designed to work closely with the database. In the same way as SQL, the LINQ is used to get and filter different data sources like collections, data sets, etc. There are many ways you can use LINQ; it is used to query object collections, ADO .Net data sets, XML Documents, Entity Framework data sets, SQL database direct, and other data sources by implementing the IQuerable interface. LINQ is like a bridge between some data and variables in the program. Every LINQ query returns objects as a result. It is good because it enables the use of an object-oriented approach to the data set. When using LINQ, you do not have to worry about converting the setups of data into objects. LINQ provides a way to query data wherever that data came from. It supports the compile-time syntax checking so that if you make a mistake while coding LINQ, the compiler will inform you immediately. LINQ allows you to query every collection such as List, Array, Enumerable *classes, etc.*

We are heading to the examples now:

```csharp
class Program
{
    static void Main(string[] args)
    {
        List<Car> cars = new List<Car>()
        {
            new Car(100, true, false),
            new Car(200, false, true),
            new Car(300, true, true),
            new Car(400, false, false),
            new Car(500, false, true)
        };

        var carsThatAreAutomatic = cars.Where(s => s.IsAutomatic == true).ToList();
        carsThatAreAutomatic.ForEach(carThatIsAutomatic =>
        {
            Console.WriteLine($"The car with the trunk capacity of {carThatIsAutomatic.TrunkCapacity} is automatic");
        });
    }
}

class Car
{
    public int TrunkCapacity { get; }
    public bool IsAutomatic { get; }
    public bool IsTurboCharged { get; }
    public Car(int _TruckCapacity, bool _IsAutomatic, bool _isTurboCharged)
    {
        TrunkCapacity = _TruckCapacity;
        IsAutomatic = _IsAutomatic;
        IsTurboCharged = _isTurboCharged;
    }
}
```

In this example, we have one class named Car. In the Car class, there are three properties, which are the TrunkCapacity, IsAutomatic, and IsTurboCharged. Every property represents read-only property, and their initial state is set in the Car class constructor. The constructor takes three parameters and assigns their values to the properties mentioned above. In the Main method, we are creating a list of Car objects. The list of Car objects is populated in the declaration part, creating five Car objects with different values for the TrunkCapacity, IsAutomatic, and IsTurboCharged properties. After that, we would filter the list and take only cars that have automatic transmission. This will be done by using the LINQ query. We are creating an object named *carsThatAreAutomatic* and filtering the cars list with LINQ. Then we perform the *Where* LINQ extension method to get the objects that we need. How does it work? In the *Where* method, we are declaring an iterative variable that will be used for filtering conditions.

The variable in this example is s. So, what the filter does next is that for every s object in the cars list, it will grant us the object whose s.IsAutomatic property is equal to true. This will, in turn, create a filtered IEnumarble list. In order to make this a list of explicit type objects, we will do the simple ToList() method over the filtered IEnumerable list. This will create an object of List<Car> type and put it down to the carsThatAreAutomatic object. Now, we have the data that we need. In the end, we will iterate through the newly created collection of Car objects and print messages to the console output. For every object in the list, we will print "The car with the trunk capacity of {certain capacity} is automatic". We know this because we filtered the Cars and took only the ones who have automatic transmission. The output of this program looks like this:.

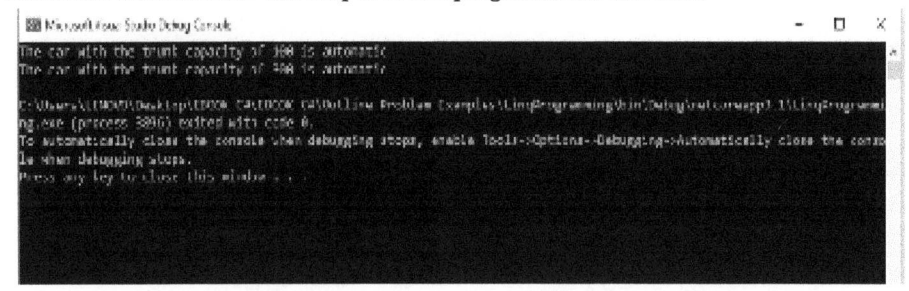

```
Microsoft Visual Studio Debug Console
The car with the trunk capacity of 100 is automatic
The car with the trunk capacity of 300 is automatic

C:\Users\LENOVO\Desktop\EBOOK CALIBOOK C#\Outline Problem Examples\LinqProgramming\bin\Debug\netcoreapp3.1\LinqProgramming.exe (process 3896) exited with code 0.
To automatically close the console when debugging stops, enable Tools->Options->Debugging->Automatically close the console when debugging stops.
Press any key to close this window . . .
```

As yu can see, there are two cars with automatic transmission and their capacity is 100 and 300 respectively. You can check that in the creation of the Car list at the beginning, as only the first and the third car have true value for an IsAutomatic property passed in the object instantiation.

Another exampleis in the next page :

C#

```csharp
class Program
{
    static void Main(string[] args)
    {
        List<Person> people = new List<Person>()
        {
            new Person(1200, "Mark", "Pent"),
            new Person(3400, "Peter", "Parker"),
            new Person(2300, "Julian", "Stones"),
            new Person(5000, "Mike", "Deen"),
            new Person(3250, "Catrin", "Burns")
        };

        var seniors = people.Where(s => s.Salary > 3000).ToList();
        var seniorSalaries = seniors.Select(x => x.Salary).ToList();
        var seniorSalariesSum = 0;
        seniorSalaries.ForEach(seniorSalary => seniorSalariesSum += seniorSalary);
        Console.WriteLine($"The sum of all senior salaries is {seniorSalariesSum}");
    }

    class Person
    {
        public int Salary { get; }
        public string FirstName { get; }
        public string LastName { get; }
        public Person(int Salary, string FirstName, string LastName)
        {
            this.Salary = Salary;
            this.FirstName = FirstName;
            this.LastName = LastName;
        }
    }
}
```

In this example, we have a class named Person. This class has three properties, which are Salary, FirstName, and LastName. Each Person class represents one employee with its basic information. The object's properties are assigned in the constructor. In the Main method, there is a creation of a list with objects of the Person class type. We have added five Person object, each with different salary amount, first name, and last name. The manager wants to filter out everyone that has a salary greater than 3000, and these people will be marked as seniors. We will do that with the LINQ query. As in the previous example, we are filtering the people list with the *Where extension method*. In this method, we are declaring that we want only the Person objects that have a Salary greater than 3000. After the retrieval is done, the ToList() method takes the filtered IEnumerable list into the Person objects list. Now we have an object that is a list of Person class types, and it contains only the people who have a Salary greater than 3000. This variable is named *seniors*. Now the manager requests that he wants the total sum of the senior salaries. We will then do that by creating a list of senior salaries and then summing all of that values into one variable and end up printing it to the standard output. The creation of the *seniorSalaries* list will be done with the execution of the Select LINQ extension method. The Select method is used when there is a need for getting only certain property values from a collection of objects. In this case, we need the Salary property value from each of the objects inside the *seniors list*. The Select method does the next: for each x object from the *seniors list, take x.Salary* value. This creates a list of IEnumerable values. After the ToList() method execution, we succeeded in creating a List<int> object that contains elements that are senior salaries. Then we created a variable in which we will store the summation of all of the senior salaries - *seniorSalariesSum* variable. After this is done, we have one more statement that will finish the work. The *ForEach* method is done over the seniorSalaries list, which will gather and sum all salaries into one variable seniorSalariesSum. In the end, the program will print the result - seniorSalariesSum value to the standard output. The output of this program will look like this:

This logic and solution to the problem could be much simplified, and the senior salary sum could be done in just one line of code. That line will look something like this:

```
var seniorSalariesSum = 0;
people.Where(s => s.Salary > 300).Select(x => x.Salary).ToList().ForEach(salary =>
seniorSalariesSum += salary);

Console.WriteLine($"The sum of all senior salaries is {seniorSalariesSum}");
```

After the Where method is being executed, the Select method would follow, after which the List will be created over which we could iterate with the ForEach extension method and do the sum calculation. This will produce the exact same result. This long statement is just split into a few smaller ones in the starting solution for better understanding. Let us close this LINQ chapter with one more example:

```csharp
class Program
{
    static void Main(string[] args)
    {
        List<Person> people = new List<Person>()
        {
            new Person( 1200, new PersonalInformation("Mark", "Pent", 123)),
            new Person( 3400, new PersonalInformation("Peter", "Parker", 155)),
            new Person( 2300, new PersonalInformation("Julian", "Stones", 133)),
            new Person( 5000, new PersonalInformation("Mike", "Deen", 143)),
            new Person( 3250, new PersonalInformation("Catrin", "Burns", 205))
        };

        var seniorsLT150 = people.Where(x => x.Salary > 3000)
                                 .Select(x => x.PersonalInformation)
                                 .Where(x => x.PersonalID < 150)
                                 .Select(x => x.FirstName).ToList();

        seniorsLT150.ForEach(seniorFirstName =>
            Console.WriteLine($"Senior name is { seniorFirstName }")
        );
    }
}

class Person
{
    public int Salary { get; }
    public PersonalInformation PersonalInformation { get; }
    public Person(int salary, PersonalInformation personalInformation)
    {
        Salary = salary;
        PersonalInformation = personalInformation;
    }
}

class PersonalInformation
{
    public string FirstName { get; }
    public string LastName { get; }
    public int PersonalID { get; }
    public PersonalInformation(string firstName, string lastName, int personalID)
    {
        FirstName = firstName;
        LastName = lastName;
        PersonalID = personalID;
    }
}
```

In this example, we have modified the Person class from the previous example. Now, the Person class contains the Salary property and PersonalInformation class type property, and both are assigned in the Person class constructor. The PersonalInformation class contains three read-only properties. These properties are FirstName, LastName, and PersonalID. FirstName and LastName are of a string data type, while the PersonalID is of an int data type. These three properties are assigned in the constructor when instantiating the object of PersonalInformation class. Now, let's jump into the Main method of the Program class. Here, we are again going to create a list of Person objects, and it is the same as in the previous example. Though, this time is going to be a bit different because of the Class modifications. We are creating five Person objects, each of them with Salary, and PersonalInformation object assigned.

For every Person, we create a PersonalInformation object to instantiate the Person object correctly. Each of the PersonalInformation objects has the first name, last name, and personal ID passed to its instantiation. This way, we created a bit more complex object that contains another class object inside. Ok, we are ready to go. The manager asks us to find every first name of an employee who is treated as a senior and has a personal ID of less than one hundred and fifty. From the previous example, we have acknowledged that the senior is the Person with a Salary greater than three thousand (3000). We are doing something similar to the previous example, but this time, the Person class has changed. It does not have the same structure as in the previous example. So, we must analyze the class structure first and then create a solution. The first step is that we must filter the Person objects from the people list that have a Salary greater than 3000. After that, we must take the PersonalInformation object from every Person object to find the first names of all the seniors. Then, when PersonalInformation objects are gathered, we must filter them and take only the ones with the PersonalID that are less than 150. When that task is done, we can finally select the FirstName property of all the seniors, make a list out of it, iterate through that list, and print the FirstName values. All of this is done in the little complex LINQ query from which we create a *seniorsLT150* list variable.

The first *Where* method creates an IEnumerable list of seniors. Then, the Select method takes PersonalInformation objects from seniors list and makes an IEnumerable list of the seniors PersonalInformation objects. The second *Where* method is working over the PersonalInformation objects from the previously created IEnumerable list, and there we will filter the PersonalInformation object that has PersonalID property less than 150. From there, we are doing the Select method, which gathers the FirstName property value for each of the seniors PersonalInformation objects that has PersonalID less than 150. In the end, we then execute the ToList() method over the final IEnumerable that we created, and the final product is the list of strings that contain the first name of every senior with a personal ID less than 150. When all of this is finally queried, we can run through the list and print those names to the console. The only senior who will meet these criteria is Mike Dean, and his first name will be printed to the standard output. The program console output is below:.

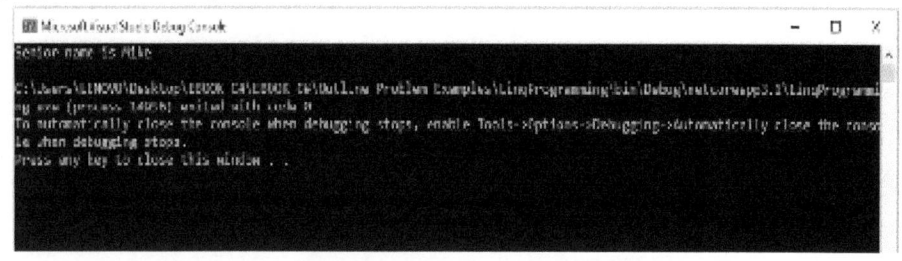

Chapter 8

Program to make decision

Decision Making
Not all problems are solved linearly, sometimes it is necessary to have to make a decision or execute specific actions when a condition is met, and others when it does not. Let us suppose that our problem is to maintain the temperature of a bucket of warm water. To do this, we can add hot or cold water. In this problem, we would need to decide what kind of water to add.

Similarly, there are many problems that we need to know a condition or make a decision on how to solve them. In C#, it is easy to achieve this since the language provides us with different tools to cope with it. We will have to use expressions, and these will be evaluated. In this case, we will use relational expressions, and logical expressions, which will be evaluated, and depending on the result of that evaluation, specific steps of the algorithm or others will be carried out. Let's start by knowing the expressions we need.

Making Choices and Decisions
While we are in the C# language, we need to take some time to work with the decision control statements or conditional statements. These are going to be the types of codes that can take the input from the user, compare it to the conditions that you set within the code, and then will make some important decisions on its own based on both of these.

There will be times when you write outcode, and you want to make sure that this program can handle some decisions on its own. You can choose to write it out so that there are specific conditions to find in the code ahead of time, and then the code can respond based on the input that the user has added into the computer.

The good news with this one is that there are a few options that you can choose from with these conditional statements. Each one is going to work in a slightly different manner based on what you are hoping to get out of the program as well, and how you would like it to respond to different inputs from the user. As we go through the different options, you will better be able to see how each one works and why it is such a great option for you to choose one of them along the way. Let's dive in and take a look at some of the main conditional statements available in the C# language, learn how each one is going to work, and explore some of the basics of coding with them as well.

A Look at the If Statement

The first kind of conditional statement that we want to take a look at here is known as our if statement. Out of all the conditional statements that we will look through, this one is going to be the most basic. It does miss out a bit on some of the functionality that we will see with these statements, but it does help us to learn more about conditional statements and can be useful in some situations as well, so it is important to learn.

When you handle this if statement, you will find that when it runs, the code will only provide us with a reply as long as the user gives us an input that matches with the conditions that are set in the code. When this does happen, then the code will be able to execute, and it will usually come up with some kind of message, depending on what the programmer puts into the code. Sometimes you will set it up to execute something else as well.

However, if the user comes through and puts in an input that does not match up with the conditions that we set up, then there will be nothing that shows up on the screen. This can cause some problems, which is why the if statement may not be used as much as some of the other options.

The good news out of all of this is that you will find, even as a beginner, that these if statements are easy to set up and work with. Let's take a look at the coding below to see how we can create one of these if statements in the C# language for our own needs: *If (x > 0)*
{
Console.Write("The value is positive.");
}
With the example that we listed above, certain things need to happen. If the value that the user put sin is higher than zero, then the program is set up to print off "The value is positive". However, if the expression turns out to be false, or the input is less than zero, then the program will ignore the whole statement that comes after Console. Write and will move on.

A Look at the If Else Statement

As we can tell pretty quickly when we work with the if statement, there will be many situations where these statements are not going to be able to help us get things done in the manner that we want. You do not want to set up some codes that are not strong enough to handle the work that you want. It is never a good idea to set up a code where the user can put in something, and then the program just ends or freezes up. And this is where we will take a look at a new conditional statement, known as the if-else statement.

The if-else statement is going to be a nice one to work with because it adds in a ton of power to the code that we are doing and writing, while still allowing us to add in more than one option and handle pretty much anything that the user will add to the program. Sometimes we will only add in one or two options, but sometimes we can use the if-else statement to handle a ton of options along the way as well.

To get to the point, we need to take a look at the syntax for the if-else statement so we get a better understanding of how this will work in the C# language. A good example of the syntax for this one will be below *If (the Boolean expression)*
{The statement/s you want to run if the result is true;
}
Else
{The statement/s you want to run if the result is false;
}
As you can already see with this one, there is going to be a lot more power that comes with the if-else statement. This is a basic option with the if-else statement, but you can add in some more lines to make it more powerful. With these, if-else statements, if the answer is false, your code will go on to the next part of the statement to see whether or not that one is true. If there are more than two options, it will keep going down the if-else statement until the answer turns out true, and then it will go and display the message that comes up.

Now that we have taken some time to talk about it, it is time to look at an example of how this will work so you can use it in your codes:
If (x > 0)
{
Console.Write("This value will be positive.");
}
Else
{
Console.Write("The value is less than or equal to zero.")
}
You are going to find that when we work with the example above, the else clause is going to be important, or at least hidden until the Boolean expression ends up being false. It is there when the process will need the else statement. But if we end up getting true value, then the first statement will be the one that will show up so we won't need the else statement at all.

What the Nested Conditional Statement is All About

Any time that we look through some of the new codes that come with C#, you should find that it is easier to work with two statements that we had above, or you can combine them to make a nested conditional statement for some of your codes as well. With a nested conditional statement, we are going to take one of the types of conditional statements above and place it inside of another one. The point of doing this is to add in some more complexity to the work that we want to see with coding and can really give us a chain of conditions when they are needed.

The main thing that we have to focus on when it comes to handling any nested conditional statement is that if there is some mistake that happens when you write out the statement, it is going to cause an error in the whole thing, and it can take some time to fix. This makes them a bit difficult to handle and work with along the way, so be careful.

The neat thing about them, though, and something that will make them pretty appealing is that they allow us to add in as many, or as few, levels as we would like. Even with this in mind, most professionals in this language will recommend that the nested statements stay around three levels or less so that they don't become too big of a mess along the way.

If you do end up going past the third level when working on these statements, then you may end up with something that is unruly and doesn't behave in the manner that you would like and may not even work. As a beginner, stick with the three levels and go up to more as you get better with the coding.

There are many codes that you are able to write that will rely on these conditional statements. They help you to make the code stronger and will ensure that we are able to handle some of the inputs that our users give to us, even if we are not able to guess ahead of time what answer the user is going to add to the program. These can base the input off any conditions that we set and will ensure that the program is going to behave in the manner that we would prefer. This makes it so much easier for us to see results and for us to make sure the program works the way that we want.

Chapter 9

Net

The .Net Framework

Programs that are written out in this kind of framework are going to be able to execute itself in the environment for the software rather than the hardware environment, known as the CLR or Common Language Runtime. The CLR is going to be an application virtual machine that will help to provide us with some different services, including security, exception handling, and memory management. Because of this, the codes that are written with this Framework is going to be known as managed code. Both the CLR and the FCL that we have talked about are going to make up the .NET framework that we are talking about.

First, the FCL is going to be important. It provides us with some of the user interface, database connectivity, data access, web application development, algorithms, network communication, and more. Programmers can produce some software when they combine the source code; they are working on with the .NET framework and some of the other available libraries.

This is a framework that is designed to work well with most new applications that we want to create on the Windows platform. In addition, Microsoft is going to produce an IDE that is specifically for the .NET software that is known as Visual Studio.

There are a lot of things that we can work with when it comes to the .NET Framework. Some methods have been developed to work not only on the Windows systems but also with some of the other operating systems, including Linux and Mac. And it can help us work with a variety of languages, especially with C#, while being portable, secure, high in performance, and good with managing the memory that we need while writing out certain codes.

In order to be able to program effectively with C#, one needs a little theoretical background knowledge, which concerns the so-called.NET Framework. That's why the. NET Framework is an integral part of this book. Before we get more involved with the. NET Framework, you should know that it is an integral part of .NET technology.

The task of a runtime environment is essentially to execute the code developed with the respective programming language. In the past, however, a separate runtime library (runtime library) usually had to be provided for each programming language - and in part also for each version - in order to be able to execute the corresponding programs. For example, C ++ the file MSVCRT.DLL and in Visual Basic 6, the file MSVBVM60.DLL. The concepts of the supplied runtime environments were and are still fundamentally different today. Cross-language support provides key benefits of the .NET Framework:

Only one framework is needed.

Different (and very different) programming languages can be used, depending on the level of knowledge and personal preference.

There is exactly one runtime environment for all programming languages.

From these advantages, further derivations can be made:

The cross-lingual integratio.
The cross-language exception handling (Exception Handling refers to how a program should react when events occur that are generally not expected, such as errors..
Improved security (security can be controlled up to the method level.
Simplified versioning (several versions can exist side by side.

Because the .NET Framework provides a cross-language runtime environment, it ensures interaction and optimal collaboration between components of different programming languages.
You now know that the .NET Framework is part of .NET technology and what it has to do. The .NET Framework again consists of three main components:
Common Language Runtime - the cross-language runtime environment
Class libraries (Learn more about classes and class libraries later)
ASP.NET - server-side applications for the World Wide Web

In this book, we will deal in detail with these three components. The common language runtime and class libraries will come up again and again throughout the book, as will ASP. NET in the context of web services. The two components - Common Language Runtime and Class Libraries - of the .NET Framework are also referred to as Base Framework.

Discover .NET
The .NET Framework is a solution to all the problems involved in the development of applications, providing great benefits not only to the developer but also to the development process. First of all, .NET allows us to work with existing code; we can use COM components, and even, in the situation where we need to use the Windows API. When the .NET program is ready, it is much easier to install on the client's computer than traditional applications since it has a strong integration between languages.

A C# programmer can easily understand the code of a Visual Basic .NET programmer, and both can program in the language they are most comfortable with. This is because all the languages that make use of .NET share the .NET libraries, so it does not matter in which language we program in, we can recognize it in any language. Next, we will know the different components of .NET: Common Language Runtime (CLR), Assembly, and Common Intermediate Language (CIL).

CLR

The first component of .NET that we will know is the Common Language Runtime, also called CLR. This is an execution program common to all languages. This program is in charge of reading the code generated by the compiler and starts its execution. It does not matter if the program was created with C#, Visual Basic .NET, or some other .NET language, the CLR reads it and executes it.

Assembly

When we have a program written in a .NET language, and we compile it, the assembly is generated. The assembly contains CIL in the compiled program and also the information about all the types that are used in the program.

CIL

The .NET programs are not compiled directly into compiler assembly code; instead, they are compiled into an intermediate language known as CIL. This language is read and executed by the runtime. The use of CIL and runtime is what gives .NET its great flexibility and its ability to be multiplatform.

The Framework of .NET has Common Language Specifications or CLS. These specifications are the guidelines that any language that you want to use .NET must meet to be able to work with the runtime. One advantage of this is that if our code complies with the CLS, we can have interoperability with other languages, for example, it is possible to create a library in C #, and a Visual Basic .NET programmer can use it without any problem.

One of the most important points of these guidelines is the CTS or Common Type System. In programming languages, when we want to save information, it is placed in a variable, the variable will have a type depending on the information to save, for example, the type can be to save an integer, the other to store a number with decimals and the other to save a phrase or word. The problem with this is that each language store information differently. Some languages store integers with 16-bit memory and other with 32-bit; even some languages, such as C and C++, do not have a type for storing strings or phrases. To solve this, the .NET Framework defines through CTS how the types will work in their environment. Any language that works with .NET must use the types as outlined in the CTS. Now that we know the basic concepts, and we can see how all this comes together.

How to Create a .NET Application

We can create a .NET application using a programming language. For this purpose, it will be C#; with this programming language, we can create the source code of the program (instructions that tell the program what to do).

When we have finished with our source code, then we have to use the compiler. The compiler takes the source code and creates an assembly for us. This assembly will have the equivalent of our code, but written in CIL, leading us to another advantage of .NET

- the compiler can optimize our code for the platform where we are going to use the program, that means the same program can be optimized for a mobile device, a normal PC or a server, without making any changes to it.

The .NET Framework can be run on many platforms, not just on Windows. This means that we can program on a particular platform, and if another platform has the runtime, our program will run without any problem. A .NET program developed in Windows can be executed in Linux, as long as it has the corresponding runtime.

When we want to invoke the program, then the runtime gets into the action, reads the assembly, and creates the entire environment for us. The runtime starts reading the CIL instructions from the assembly and compiles them as it reads them for the microprocessor of the computer on which the program is running; this is known as JIT or just-in-time compilation. In this way, the program advances in execution, and it is compiled; all this occurs transparently for the user.

The JIT compiler is also known as Jitter. It is part of the runtime and is very efficient if the program needs to re-run a code that has already been compiled, the Jitter instead of recompiling, runs what has already been compiled, thus improving performance and response times for the user.

For the programs that are running, the .NET Framework provides the services of memory management and garbage collection. In unmanaged languages such as C and C++, the programmer is responsible for memory management. In large programs, this can be a complicated task, which can lead to errors during program execution. Fortunately, managed languages like C#, have a model in which we, as programmers, no longer need to be responsible for memory usage. The garbage collector is in charge of removing all objects that are no longer needed when an object is no longer use the collector takes it and removes it. This frees up memory and resources.

The garbage collector works automatically for us and helps to eliminate all the memory and resource management that was necessary for Win32. In some special cases, like files, databases, or network connections are unmanaged resources. For these cases, we must explicitly indicate when they need to be destroyed.

Chapter 10

ENUM and struct

Enum

An enum (which stands for enumerated type) is a special data type that allows programmers to provide meaningful names for a set of integral constants.

To declare an enum, we use the enum keyword followed by the name of the enum. The members of the enum are enclosed in a set of curly braces and separated by commas.

An example is shown below:

enum DaysOfWee.

{
Sun, Mon, Tues, Wed, Thurs, Fri, Sat
}

Note that we do not put a semi-colon at the end of the last member.

After declaring the DaysOfWeek enum, we can declare and initialize a DaysOfWeek variable like this:

DaysOfWeek myDays = DaysOfWeek.Mon;

The name of the variable is myDays. If we write

Console.WriteLine(myDays);

we'll get,

Mon

By default, each member in the enum is assigned an integer value, starting from zero. That is, in our example, Sun is assigned a value of 0, Mon is 1, Tues is 2, and so on.

As members of an enum are essentially integers, we can cast a DaysOfWeek variable into an int and vice versa. For instance,

Console.WriteLine((int)myDays);

gives us the integer 1 while

Console.WriteLine((DaysOfWeek)1);

gives us Mon.

If you want to assign a different set of integers to your enum members, you can do the following

enum DaysOfWeekTwo

{

Sun = 5, Mon = 10, Tues, Wed, Thurs, Fri, Sat
}

Now, Sun is assigned a value of 5, and Mon is assigned 10. As we did not assign values for Tues to Sat, consecutive numbers after 10 will be assigned to them. That is Tues = 11, Wed = 12, and so on.

Of course, if you use a byte data type, you cannot do something like

enum DaysOfWeekFour : byte

{
Sun = 300, Mon, Tues, Wed, Thurs, Fri, Sat
}

as the range for byte is from 0 to 255.

There are two main reasons for using enums. The first is to improve the readability of your code. The statement

myDays = DaysOfWeek.Mon;

is more self-explanatory than the statement

myDays = 1;

The second reason is to restrict the values that a variable can take. If we have a variable that stores the days of a week, we may accidentally assign the value 10 to it. This can be prevented when we use an enum as we can only assign the pre-defined members of the enum to the variable.

Struct

Now, let's look at the struct data type.

A struct is similar to a class in many aspects. Like classes, they contain elements like properties, constructors, methods and fields and allow you to group related members into a single package so that you can manipulate them as a group.

To declare a struct, you use the struct keyword. An example is:

```
1 struct MyStruct
2 {
3 //Fields
4 private int x, y;
5 private AnotherClass myClass;
6 private Days myDays;
7
8 //Constructor
9 public MyStruct(int a, int b, int c)
10 {
11 myClass = new AnotherClass();
12 myClass.number = a;
13 x = b;
14 y = c;
15 myDays = Days.Mon;
16 }
```

```
17
18 //Method
19 public void PrintStatement()
20 {
21 Console.WriteLine("x = {0}, y = {1}, myDays = {2}", x, y, myDays);
22 }
23 }
24
25 class AnotherClas.

26 {
27 public int number;
28 }
29
30 enum Days { Mon, Tues, Wed }
```

The struct is declared from lines 1 to 23. On line 4, we declared two private int fields for the struct. On line 5, we declared another private field called myClass. This field is an instance of the class AnotherClass. On line 6, we declared an enum variable myDays. The two fields (myClass and myDays) are specially included in this example to demonstrate how we can include a class instance and an enum variable as the fields of a struct. Structs (and classes) can contain enum variables and instances of other structs and classes as fields.

After declaring the fields, we declared the constructor for the struct (lines 9 to 16), followed by a method to print the values of x, y, and myDays. (lines 19 to 22).

After declaring the struct, we declared the class AnotherClass on lines 25 to 28 and the enum Days on line 30. In this example, we declared the class and enum outside the struct myStruct. However, we can declare the enum or class inside the struct itself. An enum, struct, or class can be nested inside another struct or class. We'll look at an example of an enum declared inside a class when we work through the project at the end of the book.

To use the struct above, we can add the following code to our Main() method:

MyStruct example = new MyStruct(2, 3, 5);
example.PrintStatement();

If we run the code, we'll get

x = 3, y = 5, myDays = Mon

There are two main differences between a struct and a class. Firstly, the struct data type does not support inheritance. Hence you cannot derive one struct from another. However, a struct can implement an interface. The way to do it is identical to how it is done with classes. Refer to Chapter 8 for more information.

The second difference between structs and classes is that structs are value types while classes are reference types.

For a complete list of differences between a struct and a class, check out the following page:

https://msdn.microsoft.com/en-us/library/saxz13w4.aspx

Chapter 11

Common mistake and how to avoid them

This guidebook has taken some time to look through all of the different parts of coding that we need to know to utilize the C# language to write some of our codes.

There are a lot of different aspects that we need to keep in mind when it comes to working on this language, but when we can put it all together, we will find that we can create some great codes and programs in the process.

Even though many of the modern coding languages that are out there have been designed to make programming easier for those who are beginners, there are still a number of challenges that are going to show up when you first get started.

One of the things that a beginner has to remember when they go through this process is that they do need to practice, and they need to learn from their mistakes and from some of the problems that show up in their coding.

Even taking a break is often enough to help you get going on this and can ensure that you won't burn out and have trouble with the process later on.

Some of the mistakes you should avoid includes:

Not Getting enough Practice
There are a ton of codes and suggestions and things to learn about in this guidebook, but if you don't actually open up the compiler with C# and try some of it out, you are never going to gain the skills and more that you need to make this work for you.

That is why the first rule that we need to follow here is to play around with the code and get as much practice as possible.

With any of the new subjects that we want to explore and learn about in any coding language, the sooner that we can get our hands dirty and start messing around with the code, the faster that we can learn some of the concepts that are there.

You can't go through and read the information without using it a bit, and then expect that you will remember that information and be able to utilize the code for your programs.

You have to mess with the code and see how it works.

Now, you will find that the best place for us to get started with this tip is just to open up our C# compiler and start to do some of the codings that we want right away.

Take some of the examples that are in this guidebook and just mess around with them a bit.

Even just typing them into the compiler to start with is a good step in the right direction and will ensure that we can get some practice.

You can then work and explore from there to get the right results.

Ignoring the Fundamentals

Even though it sometimes seems like the fundamentals are going to be too basic to work with, and you may feel like you should just race through them without a thought, it is still important to spend our time learning about how this work.

As easy as they are, they are really important to work with as well.

The better that we can work with these fundamentals, the easier it is for us to start mastering some of the more advanced stuff that is going to show up.

Those programmers who try to get into a programming language and then rush through the beginning parts and do not spend the proper time on some of the fundamentals are going to be the first ones who get stuck when they need to make the transition int some of the advanced material that will come later.

So, before you miss out on some of the first classes that we need here and skip through some of the basics that are important in all of this, make sure to learn more about the fundamentals and what we can do with it along the way.

Not Writing it Out

It is normal to want to get started with programming and to open up the compiler and start coding.

And while this is one of the methods to use, sometimes trying it out in a different manner, and writing it by hand rather than trying to type it in all of the time is going to be the trick that will help us to get this done.

Bringing out the pen and paper to write out our codes will help.

There are a ton of advances to the computers out there, and they're a lot of benefits to using it.

But sometimes the best way to learn something new is actually to write it out and work from there.

Whether you decide to use some scrap paper, a notebook, a whiteboard, or another option, taking the time to code out everything by hand is going to take more time to work with.

There is the requirement to use more caution, precision, and even intent behind all of the lines of code that you try to write out.

You also are not able to check out the code when writing it like you can on the computer, which forces you to pay more attention.

This method is going to take more time and be really consuming when it is time to get things done.

But it is going to be a great method that helps us to go through and become a better developer in the process.

And if you plan to use this through college or for a new job, being able to write out the codes that you are using and utilize this for your needs is going to be so important to the success that you are able to see as well.

The more time and practice that you are able to give to writing out some of your codes by hand, the better you will get at understanding the coding and all that you can do with it.

It forces the programmer to slow down and actually focus on what they want the codes to do.

And then you can use this as a way to catch your own mistakes and learn what works the best for you and what does not.

Not Asking for Help

When we first get started with writing out codes in this manner, it is possible that you have a ton of grand ideas of how things are going to work and how you will become the greatest code writer of all time.

You may believe that nothing is going to go wrong with this and that you can handle it all in no time at all.

And one of the biggest misconceptions that are going to show up with the work that we want to do here is that we assume we really don't need to start any kind of help with our coding at all.

While it would be awesome to get started with coding in any language, even C# on our own without any help, we have to face reality a bit and remember that we are going to learn in a faster and more efficient manner when we have the right kind of peer feedback and mentors to help us out with this.

What may seem like an impossible bug to work with or a topic that seems like it is unlearnable when you do it on your own, you will find that when someone with more experience steps in and can help out, things get a lot easier, and you can actually learn something new.

Whether it is online where you ask for help, or you are able to find someone who is able to come to you in person and offer advice, you should never be scared to ask someone for help.

All of the programmers who are more advanced right now were at one time in the same place as you are now.

And most developers are going to love that they actually get a chance to code again, and will be more than willing to help you out with some of the codings that you would like help with.

Of course, we need to be careful with this and not take it for advantage at all through the process.

The best rule to follow here is to never take more than 20 minutes getting someone to help you figure out something with your code.

And you should not ask for help without spending at least 20 minutes on it ahead of time, trying to learn how it works and what you are able to do with this as well.

Not Taking Breaks
There are going to be times when you are writing out some of the codes that you want to do in the C# language, and then you get stuck on something.

You spend some time on it, but that just seems to make the whole thing worse.

You keep working at it and working at it, effectively making the problem worse, or not being able to find the problem at all, and your frustration levels keep going up.

You want to be able to fix the problem and get on with your coding, but you just get angrier, and the codes get messier in the process.

When this happens, and this is something that can happen to everyone, we must take the time to take a break.

This is probably the last thing that we want to worry about when it comes to working with the C# language and some of our codes, but it is going to allow us to take a break from the problem and get some fresh air, or at least do something else for some time.

And often, after you take a break and then come back to it, you will find that the problem, which seemed impossible at the start, is actually really easy to fix.

No one wants to give up when they have put in so much time and effort to this process, but in the long run, it can cut down on the frustration and will ensure that you can fix up your code in no time.

Ignoring the C# Community

One of the neat things that we are going to find when it is time to do our coding in the C# language and more, is that there is a large community of different programmers and developers along the way who are able to help you out.

These communities are going to include a lot of programmers who have been in all stages of the process.

Some are beginners, some have been in the game for a bit, and some are more advanced.

This is great news for you because it allows us a chance to go through and learn a lot of things.

We can ask questions of those who are in the community.

We can find a lot of the codes that we need to help us work on a variety of programs and learn something new.

And we are able to talk to others about the programming language and meet some great people while asking a lot of questions along the way as well.

It is a good idea for you to go through and make sure that you are able to really find the community that works for you.

There are a ton of these communities found online, and we just need to make sure that we find the one that seems the best for us.

This is a great place to resort back to when there are some problems with your code or when you would like to be able to get something figured out that is not working out that well to start with for your code.

Failure to Utilize the Sample Codes Provided

While this guidebook took some time to show you a few samples of code to show how all of the different types of coding are supposed to work, it is not enough to build up your understanding just to look over the code.

To help develop a true understanding, and you need to take some time to run and tinker with the code to see what it can do.

The more times that you are able to spend working on the code, the better off you will be along the way.

With the addition of things like instructions and comments, the sample codes that you are going to work with are packaged in a manner that is digestible in an easy manner by the reader.

But you will find that sometimes, these are hard to replicate from scratch.

Reading is not going to be the same thing as understanding, and actually going through the process of writing out some of the code on your own, and then running it, is going to make sure that you can learn how to code much faster than before.
The more that we are able to spend writing and practicing some of the codes that we find, and then testing these codes out as well, and the more that we can tinker with these as well, the easier the coding language is going to be along the way.

This is a great way to make sure that we are able to learn what is going on, what will not work when we make changes and a lot more in the process.

Just reading the code may work in some cases, and it seems like it is the better option to work with, but it is not going to show us the best way to be efficient in order to actually create your own programs.

Working with the C# language is a great choice to make.

There are a lot of options that we are able to focus on, and it is a good option for most of the programs that we want to be able to create.

Even as someone who is brand new to the world of programming and is not used to doing any kind of coding, I will find that the C# language is a good option to work with.

When you are ready to get started with the process of learning a new coding language, you will find that avoiding these mistakes you will able to get started on the right foot.

Conclusion

If you have decided that the C# language is the right one for you to work with, then you are in good company.

There are many programmers throughout the world who are going to use this language, and now that you are done with this guidebook, you can join the ranks now as well.

There are so much knowledge and information inside of this guidebook that we are able to work with, and you will find that this is going to be one of the best options that we are able to utilize when it comes time to work with the C# language.

Contributing to C # 's ease of use is the reduction of some C ++ features, and no longer macros, templates, and multiple inheritances. Especially for enterprise developers, the above features will only produce more trouble than benefits. New features that make programming easier are strict type safety, version control, garbage collect, and more. The goal of all these functions is to develop component-oriented software.

Your effort to learn C # is a big investment because C # is designed for the primary language of writing NGWS applications. You will find a lot of functions that you can achieve or laboriously implement in C ++, which are just some basic functions in C #. For enterprise-level programming languages, new financial data types are popular. You use a new decimal data type that is dedicated to financial calculations. If you don't like this simple ready-made type, you can easily create a new data type based on the special needs of your application.

C # provides you with convenient functions such as garbage collection, type safety, version control, and so on. The only "cost" is that code operations are type-safe by default, and pointers are not allowed. It's all about type safety. However, if you need pointers, you can still use them with non-secure codes, and you cannot have column sets when calling non-secure codes.

C# is an easy programming language, making it good even to the beginners in programming. One only needs to setup the environment and start writing and running their C# code. You can use C# alone to develop a complete computer application.

As a beginner, you will accomplish amazing goals with this guidebook and gain valuable knowledge in learning how to work with the C# language. We have explored many different topics that come with this language, such as how to work with conditional statements, how to work with classes, and what the operators are all about, and so much more. This guidebook will help you to go through the C# language so that you can learn to write your own codes.

If you feel like you're ready to learn a new coding language and get started on programming, C# is a great one to go with. Make sure to check out this guidebook to learn all the basics that you need to know to create really useful codes in no time at all.

Good luck and happy programming.